SCIENCE EXPERIENCES FOR
THE EARLY CHILDHOOD YEARS
An Integrated Approach

SIXTH EDITION

JEAN D. HARLAN

MARY S. RIVKIN
University of Maryland—Baltimore

Merrill,
an imprint of Prentice Hall
Englewood Cliffs, New Jersey   Columbus, Ohio

Library of Congress Cataloging-in-Publication Data
Harlan, Jean Durgin.
      Science experiences for the early childhood years : an integrated
approach / Jean D. Harlan, Mary S. Rivkin. — 6th ed.
            p.   cm.
      Includes bibliographical references (p.   ) and index.
      ISBN 0-02-350180-4 (pbk.)
      1. Science—Study and teaching (Early childhood)   I. Rivkin, Mary
S. II. Title
LB1139.5.S35H37   1996
372.3049—dc20   95-17078
                                                                      CIP

Cover photo: Richard Hutchings/Photo Edit
Editor: Bradley J. Potthoff
Production Editor: Sheryl Glicker Langner
Design Coordinator: Julia Zonneveld Van Hook
Cover Designer: Brian Deep
Production Manager: Laura Messerly

This book was set in Palatino by Carlisle Communications, Ltd. and was printed and
bound by Quebecor Printing/Book Press. The cover was printed by Phoenix Color Corp.

© 1996 by Prentice-Hall, Inc.
A Simon & Schuster Company
Englewood Cliffs, New Jersey 07632

Printed in the United States of America

10 9 8 7 6 5 4 3 2

ISBN: 0-02-350180-4

Prentice-Hall International (UK) Limited, *London*
Prentice-Hall of Australia Pty. Limited, *Sydney*
Prentice-Hall of Canada, Inc., *Toronto*
Prentice-Hall Hispanoamericana, S. A., *Mexico*
Prentice-Hall of India Private Limited, *New Delhi*
Prentice-Hall of Japan, Inc., *Toyko*
Simon & Schuster Asia Pte. Ltd., *Singapore*
Editora Prentice-Hall do Brasil, Ltda., *Rio de Janeiro*

Preface

As education reforms move at a glacial pace toward the National Education Goals for the year 2000, conflicting social demands on schools escalate briskly. These pressures can increase teachers' responsibilities and reduce their time to actually connect meaningfully with children. We believe that our integrated approach to guiding early childhood science continues to reduce the fragmentation of teachers' efforts and childrens' experiences. Weaving curricular areas together helps create more harmonious learning and teaching. This approach offers children lasting, satisfying, and relevant understanding of the world around them. It gives them a strong foundation and an appetite for future science learning, as well as the sense of competence needed to succeed with it.

It seems more important than ever to acknowledge the socioemotional underpinnings of cognitive learning when planning activities for children. Growing up has always been challenging, but today's school-age children seem increasingly stressed, as the sharp increase in prevalence of attention disorders suggests. More than ever, teachers need to give priority to maintaining a warm, nurturing, child-sensitive classroom climate.

This sixth edition continues to report current cognitive research findings that emphasize the centrality of the learner's existing understanding and the importance of the teacher's role in sensitively advancing those prior understandings. This edition also continues to appeal to children's imaginations and to address the emotional aspects of science education.

A new chapter on the environment is included to help children understand the connections within our environment. The activities also help integrate those from the chapters on air, water, plants, and animals.

The nutrition experiences have been redesigned around the Food Guide Pyramid. New activities emphasize the importance of grains in a healthy diet.

Responding to increased national awareness of deficiencies in math understanding, the math sections in each chapter have been strengthened. Math skills continue to be integral to many of the activities. Counting, graphing, and classifying are essential science tools.

The assessment section has been expanded, again emphasizing the value of ongoing teacher observations and asserting the limitations of behavioral objectives. Suggestions are made for involving children in their own evaluation.

As usual, this new edition updates references and annotated bibliographic sections. They reflect the outpouring of new science trade books that encourages us.

Revisions throughout this edition have been leavened and informed by the insights and enthusiasm of coauthor Dr. Mary Rivkin. The breadth and depth of her invaluable expertise range from preschool, elementary school, and university teaching to active involvement with professional science and early childhood education groups at the national level, making her the best addition to this edition.

The intent of this book is twofold: First, to inspire teachers to help children learn what they ardently wish to know—how the world really works. Second, and equally important, to inspire the kind of teaching envisioned, ironically, by the 17th-century philosopher, John Locke. He wrote that he had always imagined that:

> Children . . . might be brought to a desire to be taught, if only learning were proposed to them as a thing of delight and recreation, not as a business or task.

ACKNOWLEDGMENTS

Bringing vitality and authenticity to a college textbook of this sort requires ongoing interactions with real children in real classrooms. Since I now devote major time to the practice of psychology with adults and children, my classroom contacts have diminished. Clearly, I needed a seasoned, knowledgeable front-lines partner in order to write a fresh, current sixth edition of this book. Fortunately, I was introduced to Mary Rivkin by Polly Greenburg, the Journal Editor of the National Association for the Education of Young Children. Our views on science education and our temperaments meshed. The revision was undertaken and accomplished together. My other great need was for patient computer assistance. Thankfully, this was supplied generously by Karin Harmon and Richard Victor.

Sally Warner Bahrke and Gustave Carlson provided new illustrations, and Bonnie Bahrke supplied computer graphics. The staff of the Oak Creek, Wisconsin, Public Library were unfailingly helpful in tracking down elusive resources. Helpful comments were supplied by the following reviewers: Cecelia Benelli, Western Illinois University; Catherine Blount, Cuyahoga Community College; Karen Colleran, Pierce College; R. Eleanor Duff, University of South Carolina; Blythe Hinitz, Trenton State College; and Colleen K. Randel, University of Texas at Tyler. Over the years of its publication, this book has been enlivened by the drawings of my dear friend, Anne Clark Culbert, whose memory I honor here.

Finally, for the richness they have given to my life I thank my children: Betsy Bales, Dr. Anne Strohm, Dr. John Harlan, Susan Borghese, and Julie Harlan-Schneider; and my wonderful grandchildren: Kate, Rachel, Liz, Christopher, Lauren, Nina, Sophia, and Laura. I am grateful to all of you.—J.H.

For field-testing the activities and experiences in the Environment chapter, thanks to the University of Maryland, Baltimore County students in my early childhood math and science processes seminars: Kim Arbaugh, Bonnie Bahr, Lisa Bienenstock, Laura Bradley, Chris Butters, Marge Cameron, Denine Carter, Carrie Colvard, Laura Danos, Kathy Fitzpatrick, Janie Gooch, Terry Hanford, Hwaida Hassenein, Christine Hilker, Christine Hoover, Catherine Howanstein, Michelle

Johnson, Angela Kady, Kathleen Kellam, Janna Kish, Jennifer Lockwood, Debbie Manville, Mary Mason, Janelle McIntyre, Amy Meyers, Heather Mitchell, Shannon Moe, Tina Morest, Stacie Mueller, Devin Shannon, Helene Siegert, Lara Silverstein, Valerie Sinclair, Michelle Strasnick, Jill Sturgis, Len Taylor, Gina Tognocchi, Dawn Woods, and Peggy Yockey.

Thanks to my family for help and encouragement: Steve, Jesse, Caroline, Gustave, Rob, Ina, and Susan. Special thanks to Polly Greenberg for opportunities, support, and spirit.—M.R.

Contents

PART ONE

THE RATIONALE

1

An Integrated Approach to Science Learning

AFFECTIVE COMPONENTS OF LEARNING

"Look, Teacher, my hand!" Jimmy spoke! At last, Jimmy had broken through the stoic silence of his first, anxious month in Head Start. Crouched on the carpeted floor to blot up his juice spill, Jimmy was transfixed with wonder at the imprint of his hand on a paper towel. The surprise of encountering the familiar in an unexpected circumstance captured him both intellectually and emotionally. As Jimmy questioned his teacher and experimented with fresh towels, he chattered about his discovery with other children. His own irrepressible curiosity had finally drawn him into the circle of learners at our center.

How can a 4-year-old like Jimmy focus with such intensity on the mundane experience of absorbency? Why, for that matter, do children want to know why? One part of the answer is that understanding the environment by interacting with it is the natural work of young children. From birth onward, infants use and slowly coordinate their reflexes and perceptions into information-gathering resources. The early looking at, grasping, and mouthing of objects gradually build into specific ways to recognize and handle different materials. Simple discoveries of cause and effect allow babies to solve playpen problems. If children's curiosity and exploring are encouraged, their probing for meaning broadens as language and locomotion develop. By giving children opportunities and support, their inherent need to know continues to deepen and serves as part of their motivation to engage in more complex investigations such as Jimmy's.

When the human desire for understanding the world is organized into careful ways of collecting, testing, and sharing information, it is called *science.* Indeed, the word *science* is derived from the Latin root *scire,* meaning "to know." When we offer intriguing science experiences to young children, we nourish their natural, human capacity to know. If we do this with sensitivity to their interests, nature, and needs, we release and enhance the powerful affective component of knowing and learning.

The affective component of knowing and learning is a complex web of interrelated facets, including curiosity, emotional responses to life experiences, and the self-esteem that stems from becoming competent. The way children feel about themselves and their world influences their curiosity. There is an energizing, reciprocal link between finding out and self-esteem, between feelings of mastery of the newly learned and the desire to know more. Anxious concerns about unexplained or frightening events may lure children into finding answers. Pleasure in discovering awesome, delightful, or comforting features of the environment also furthers affective and cognitive growth. This chapter will consider some effects of these interrelationships, their biological basis, and multimodal ways to use these interrelationships and biological givens to enhance science learning.

Curiosity and Emotions

Many of the existing theories about curiosity view it not as an emotion in itself, but as an *affect:* a mental state that influences emotions. Researchers have also demonstrated that emotions, in turn, have a strong influence on curiosity. In classic studies, Harlow (1958) analyzed the spontaneous curiosity and exploratory behavior of infant monkeys, and Ainsworth and Bell (1970) observed that of human infants. Both investigators confirmed that feelings of security enable curiosity and exploration, while insecurity and strong fear can interrupt and even paralyze curiosity. Other studies by ethologists add that, while overwhelming fear prevents exploring, uncertainty tinged with a bit of fear of the unknown seems to stimulate curiosity and exploring.

Pleasure and curiosity combine into lasting learning.

Emotions and Learning Interest

It is easy to observe the lasting interest in learning that can result from a warm firsthand experience. This was the case when a group of children in a day-care center unexpectedly discovered an untended nest of young rabbits in their play yard. The children became involved in the efforts to keep the babies fed and sheltered until they were old enough to leave the nest. For many months, the children spoke about "their" rabbits. They responded eagerly to new information about rearing animal young. The pleasure and pride they gained from nurturing the rabbits provided a positive emotional link to the new learning. This example demonstrates one way that mutually enhancing cognitive/affective processes lead to meaningful learning.

Negative emotions can also stimulate the need to know. This seemed to be true of Jake. One of the smallest children in our class, Jake endured regular pummelings by his older brother at home. The boys' parents did not intervene, since following their "boys will be boys" beliefs required no action. But Jake brought his resentment to school, moving warily among his male classmates with clenched fists at the ready. Jake was fascinated by our gravity experiences. As often as he could coax a partner, Jake claimed the playground teeter-totter where he could be in control. He could keep a bigger boy dangling above the ground, or be in charge of creating a balance between themselves. In the classroom he focused long, solitary attention on using the balance scale, and on finding ways to balance pennies on a ruler placed on a half-circle block fulcrum. Each time he mastered his self-created balance challenges, he exulted, "Now they're even!"

Occasionally the connection between a child's anxious concern and his or her eagerness to find answers is clearly apparent. For example, the mother of 4-year-old Matthew asked for advice on how to help her son get to sleep at night. He was unable to shut his eyes until he had counted all the stars he could see from his window. He needed to count them over and over to make sure they were still all in place. That provided reassurance for him that a star wouldn't fall on him while he was sleeping.

In response to his concern, a series of experiences was provided to build understanding about why things fall to earth, or in this case, why they don't fall. Matthew enjoyed some tangible explorations of the effects of magnetism, an invisible force that could, nonetheless, be felt in the hands of a fascinated boy. After many experiences with magnets, Matthew experimented with some simple effects of the greater invisible force of gravity. He found that the hardest toss by a strong boy couldn't send a ball beyond the pull of the earth's gravity. Even he himself always came back to earth after his highest leaps. Matthew helped build a carton spaceship to play at being an astronaut, floating so far from earth that its gravity could not pull him down. He frequently asked to be read a story about astronauts who traveled 3 days to reach the moon, because it was so far away. He always chimed in at the part that told how stars were even farther away from earth than the moon.

The day finally came when Matthew's mother reported that it was now easier for him to get to sleep. He only had to do one quick scan of the stars before climbing into bed. He explained confidently to his parents that earth's gravity is-n't strong enough to pull down a star. In the years since then, Matthew's fascina-tion with astronomy has continued. He and his family have made several visits to an observatory and a planetarium in a distant city, and he now owns a small tele-scope. The story of Tony in Selma Fraiberg's classic book *The Magic Years* (1950) is a similar example of how a child's fear led to an enduring interest in science.

A reviewer of the first edition of this textbook in 1975 was startled by its then radical, affective-cognitive orientation. He scribbled across one page (with some emotion, apparently), "What have feelings got to do with anything? This is a book about science!" Since then the reality of the coexistence of emotions and thought has been confirmed as biological fact. Thoughts and feelings occur in separate ar-eas of the brain; however, these areas are intricately connected by two communi-cation systems: electrical (nerve pathways) and chemical (neurotransmitters). Depending on their type or quantity, the neurotransmitters can either intensify or suppress memory. Mild stress and anxiety release the hormone adrenaline, which facilitates recall. However, high levels of anxiety produce excessive outpourings of adrenaline that reduce recall. Pleasurable feelings and a perception of novelty and complexity stimulate the memory-facilitating center of the brain.

Feelings are triggered by thoughts, and thoughts are affected by feelings, usually without our conscious awareness. As Restak (1991) pointed out, "it's al-most impossible to have a thought without at the same time having some feeling about it" (p. 124). It has also been well-established that emotionally significant events are retained more powerfully in memory than are emotionally muted events. Rachel Carson (1960) wrote in her influential work on nature study, *A Sense of Wonder:* "Once the emotions have been aroused, then we wish for knowledge about the object of our emotional response. Once found, it has lasting meaning." Surely the influence of feelings on what and how children learn about the world, and subsequently act toward it, is far too important to disregard.

Problem-Solving, Competence, and Self-Esteem

Science experiences have special potential for building a sturdy concept of the self as a person who can cope with problems. Martin Seligman's motivation studies (1992) have led him to conclude that the belief in one's own competence begins in infancy and develops throughout life as mastery motivation. He believes that if young children are not provided with, or are not allowed to cope with, problems that can be resolved through their own actions, a pattern of helplessness begins. Seligman feels that success which is too easily reached, as well as challenge which is too easily met, produces children with a limited capacity to cope with failure. Therefore, he suggests offering learning challenges in school that children can mea-sure themselves against, since meeting challenges helps shape a person's sense of self-worth.

According to Seligman, our self-esteem and sense of competence do not de-pend so much on whether good or bad things happen to us, but on whether we be-lieve we have some control over what happens to us. Early science experiences can provide children with some sense of control through allowing them to predict that certain things will happen; for example, "The pan of snow will change to water if we keep it indoors." Through this process, some of the confusion and uncertainty about events occurring around the child can be replaced with that awareness of predictability. The child can learn that even she can bring about some of these small occurrences. Science knowledge helps develop the only control a child can have over powerful and sometimes unpredictable natural forces. For example, children can control their attitudes through understanding the causes of an event and can control their helpless fear through training to cope safely with events such as earthquakes and violent storms. Science experiences are unique in fostering this strengthening influence on the child's personality.

LEARNING: A MULTIMODAL PROCESS

The practice of extending learning through different modes has been followed by nursery and progressive school teachers for more than half a century. Early research has shown that learning is facilitated when ideas are presented to young children in a variety of contexts (Razran, 1961). This was followed by the Nobel prize-winning research of Roger Sperry (1974) that provides further insights into why this is so.

In the late 1960s Sperry investigated brain wave patterns of patients whose right and left brain hemispheres had been separated surgically, severing the nerve pathways that otherwise joined them. Sperry found strong evidence that each half of the outer part of the brain (the cortex) processes the same information in different styles. Left-brain thinking appeared to be abstract, analytical, sequential, and logical. It used symbols—language and numbers—to think. It tended to be rational and linear, with one thought leading to the next in an orderly fashion, often resulting in convergent conclusions. Sperry's patients had conscious awareness of their left-brain activities, because that hemisphere thinks with language.

Sperry and later researchers concluded that right-brain thinking was intuitive, holistic, and more influenced by emotion than was left-brain thinking. The right brain seemed to use images and metaphors for thinking, so it appeared to think a bit faster than did the left brain. The right brain tended to serve synthesizing, creative problem-solving functions, being better able to deal with the novel and unfamiliar than was the left brain. Because the right brain seems to use images instead of language and numbers for thinking, right-brain thought doesn't enter directly into conscious awareness. It remains unconscious. There are many examples of how unconscious creative problem-solving has yielded the keys to major scientific breakthroughs, and how such thinking has inspired great works in the arts and literature. Actually, a large amount of our mental activity goes on unconsciously (Restak, 1991).

However, more recent research suggests that it is an oversimplification to think of these two hemispheric processing forms as if they functioned independently. More sophisticated brain scanning devices now make it possible to more clearly trace patterns of ongoing internal brain activity in subjects with intact brains. Such brain scans provide a picture of both sides of the brain working together as a whole. While newer evidence suggests that the various tasks may be more widely scattered through the cortex than the earlier studies had determined, the essential idea of conscious and unconscious functioning of the differing sides of the brain stands unchallenged (Buzan, 1994).

As neuroscientists continue to unfold new information about brain mechanisms, what is already understood about this incredible organ suggests effective ways to promote intellectual growth. Even after the brain has reached its mature size, intellectual growth continues to occur throughout a person's lifetime as neural connections increase between brain cells. Optimally, these neural pathways grow in size and complexity when they are stimulated by new information from the environment. The more neural pathways that are created and used, the more effective learning and thinking becomes. The resulting implications for good teaching validate the intuitive understanding of those early nursery school and progressive teachers: that many learning pathways need to be provided for youngsters to engage their enormous intellectual potential most effectively.

Two Ways of Knowing

Western cultures have tended to respect the educated mind for its store of knowledge and its orderly, logical thought processes. In other words, rational thought that can be promoted through formal teaching has had a high value. Intuition, inner wisdom, knowledge based only on practical experience, or emotion-tinged gut feelings have been less valued. Yet, we have all experienced the relief of having a name we had been unable to recall suddenly pop into mind some time later. Many of us are aware of thinking on two different levels simultaneously when we listen to someone and, at the same time, evaluate what is being said, or evaluate the speaker. Or we may have struggled to resolve a stubborn problem and then given up. Moments or hours later, a fresh insight came into conscious awareness, seemingly "out of the blue," and the problem was resolved! A creative solution was in progress in one side of the brain even after the other side stopped fretting about it. Even though we have had such experiences, we may be uncomfortable with the idea that part of our mental processes could be occurring without our conscious awareness. We feel more in control when we rely on our conscious thinking processes.

Nonetheless, recent neurological evidence seems to bear out what psychological research and various psychotherapy theories have been pointing toward for many years (Epstein, 1994): Humans have two valid, but different, intellectual processes, both capable of working together for the most effective use of the fantastic brain. We can identify these two kinds of knowledge as:

1. *Rational knowledge,* which is verbal, logical, and systematic, probably originating in the left brain. We tend to regard scientific and mathmatical thinking as primarily rational thought. We are consciously aware of thinking rationally.

2. *Intuitive knowledge,* which is a rapid, nonverbal synthesis of emotions and perceptions from both sides of the brain. Now regarded as right-brain processing,

its enlightening, creative insights come into our conscious awareness during periods when the left brain is relaxed, or focusing on other thoughts. We are not consciously aware of using intuitive knowledge, since it uses images and metaphors in a shortcut process. Apparently this form of processing continues during periods of sleep and rest for the rational left brain. We recognize intuitive knowledge only when it comes into our conscious awareness.

Both forms of knowledge can be accurate, and each form can enhance the other. Traditional, formal teaching styles are directed toward rational knowledge. Intuitive knowledge has been distrusted and dismissed in the classroom as being less verifiable and too rapid to be explained by logical reasoning. However, we can no longer afford to overlook the learning enhancement possible by recognizing and intentionally engaging the creative problem-solving qualities of intuitive thinking. We need to stimulate it and combine its use with rational thinking for optimal teaching.

Jonas Salk revealed that he began early in life to imagine himself as the object he was studying. As a scientist, he would play with several mental scenarios for a problem to acquire new ideas, then would develop his experiments accordingly. He said, "I was aware that I was using my imagination to explore reality" (Salk, 1994).

In the realm of scientific thinking, intuitive insights can be essential both for identifying problems to be solved and for having hunches about how to logically pursue a solution (Gardner, 1993, p. 148). For centuries there has been evidence of the role intuition and imagery have played to inspire great scientific discoveries and masterworks in the arts and humanities. Groundbreaking discoveries like the production of insulin and the formulation of the first model of atomic structure have come to scientists during periods of relaxation and sleep.

We can stimulate optimal learning when we intentionally provide multiple pathways or modalities for processing information. Providing children information in different ways allows children to integrate information in the uniquely preferred styles identified in Gardner's forms of intelligence (see page 64). We provide for this needed integration when we offer children creative ways to express what they have learned. We also encourage integration when we build in time to give logical thinking a rest, so children can reinforce learnings through imagery.

We encourage intuitive/creative thinking when we respect spontaneous demonstrations of children's intuitive and rational thought working together. Andrew offered a charming example of this during a class discussion of the story *Little Bear Goes to the Moon* by Minarik. The children were speculating about what would happen when Little Bear, pretending to fly to the moon, actually jumped from a tree stump. Andrew's prompt response was, "See, air and gravity are fighting. Air wants to hold Little Bear up, so he will float. But he's too heavy, so gravity will win and pull him down." Andrew recalled a concept about air he had acquired 5 months earlier. He integrated this idea with his hunch about gravity's effect on Little Bear, and expressed his ideas in a metaphor familiar to this aggressive 5-year-old.

AN INTEGRATED LEARNING FRAMEWORK

Primary Integrating Activities

When we integrate science experiences with other curriculum areas, we help children enhance their mental performance. As we enrich the range of connections and relationships among different styles of absorbing and applying information, children form more intricate neural pathways in their brains, and concept retention increases. Children with varying forms and degrees of intellectual bent can find emotionally satisfying, meaningful paths to learning. When we let children see otherwise abstract concepts actually functioning in their familiar world, another significant learning link is forged. For these reasons, a variety of enriching extensions are suggested for each of the major science topics found in Part 2 of this book. Each of the learning modes described in the following paragraphs draws upon one or more of the multiple forms of intelligence identified by Howard Gardner in his book *Frames of Mind* (1992). (See page 17.) The effect is diagrammed in Figure 1–1.

Math activities are an integral part of all science because they provide ways to quantify and record observations. Some of the suggested math activities are necessary parts of a science experience. Others use science themes to provide a new context for using math skills. Both quantifying and numerical reasoning rely on the *logical-mathematical intelligence* identified by Gardner.

Music can provide several cues for strengthening science understanding. Melody can evoke positive feelings about these concepts. Lyrics can use actual or

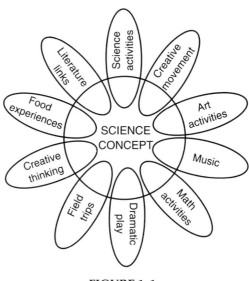

FIGURE 1–1

metaphoric ideas to strengthen recall. Rhythm reinforces song ideas through repetitive patterns. Hearing, in itself, is a sensory system that evokes strong memories. As tunes run through children's minds, they are reminded of the ideas expressed in the lyrics. Catchy melodies, lyrics, and rhythms are routinely used to remind consumers to buy products, because they are such powerful memory cues. Songs and ballads were used to educate for centuries before books and schools existed. Tapping into *musical intelligence* can make learning easier and more durable.

Literature links extend science concepts by associating them with fresh language and images, both metaphorically and as narrative. Whether incidentally embedded in fiction or introduced as focal themes of science-based stories and poems, science ideas register emotionally in children's minds as they encounter familiar knowledge in new situations. Encouraging children to write or tell their own stories and poems promotes the creative synthesis of fact, fantasy, and feelings, which uses *linguistic intelligence.* Preschool children also enjoy fingerplays, which incorporate the *bodily-kinesthetic form of intelligence* into science learning.

Art activities are suggested in the text to stimulate intuitive, creative expressions of children's own ideas. They are deliberately open-ended to encourage personal interpretations of science events. As children draw, paint, and model to represent what they have learned, they engage both *spatial* and *bodily-kinesthetic intelligence.* Some of the art suggestions incorporate materials used in the science experiences. This encourages divergent thinking, because children invent new ways to use the materials. Craft projects that require children to follow specific directions to achieve a certain end-result product are not suggested as art activities.

Dramatic play situations afford young children chances to test out and apply science ideas imaginatively. The text suggests two forms of play: creative drama ideas for impromptu guided dramatization of known stories, and spontaneous play themes for which a few props are supplied to stimulate children's own play ideas. Barbara Biber (1979) reminded us that dramatic play has been considered a valid form of learning since the pioneering days of nursery school nearly 75 years ago: "It was seen as a means of deepening insights, integrating knowledge, and finding identification on a personal level." Later research showed that play and science are complementary aspects of problem-solving. Science gives structure to activities, while play encourages creative behaviors and positive attitudes toward solving problems (Severide & Pizzini, 1984). Note, however, that any play loses its value as relaxed, pleasurable activity if adults hover closely, ever ready to capitalize on "teachable moments." Dramatic play based on a science theme can draw upon *linguistic, spatial, bodily-kinesthetic,* and *interpersonal forms of intelligence.*

Creative movement is a joyful, relaxing way to increase conceptual understanding and strengthen retention of information. Physical encoding (forming mental symbols) takes place as abstract ideas are intuitively translated into concrete physical movements of the body. Spontaneous expression through movement calls forth both *spatial* and *bodily-kinesthetic forms of intelligence.*

Creative thinking is encouraged through open-ended strategies that reframe in new ways what has already been learned. Using visualization and imagination, science concepts can be tested and clarified by reversing events, taking ideas apart to create fantasy solutions, and looking at ideas from new perspectives. Such activities encourage flexibility in shifting between rational and intuitive styles of thinking, and can stimulate new interest in a science topic. Creative thinking activities draw upon *intrapersonal intelligence.*

Food experiences use taste and smell to strengthen the recall of concepts. The pleasure of being involved in preparing or sampling good things to eat provides a heightened emotional state for lasting memories. All of us have vivid connections between particular foods and the memories of the feelings we associated with those foods. Edible science experiences can strengthen concept retention by using *bodily-kinesthetic intelligence.*

Field trips add relevance and tend to validate the science information learned in school. Children are proud to recognize that what they know from the classroom has significance in the real world. It is important to invite visitors into the classroom who are involved in work that applies science concepts that students have learned, thus bringing the field to the school. Classroom windows sometimes become impromptu resources for verifying science learnings. For example, the unexpected appearance of a squirrel on the window ledge, or the activity of a street repair crew can become a welcome illustration instead of a distraction. Depending on the nature of the field trip, many of the forms of intelligence can be drawn upon.

This integrated approach to science education weaves physical, sensory, and emotional activities into the total learning process. It encourages the use of both rational and intuitive styles of thought. This approach is also consistent with the recommendations of the National Center for Improving Science Education (Bybee, Buchwald, Crissman, Heil, Kuerbis, Matsumoto, & McInerney, 1989). They suggest that themes and topics for science should

- build upon children's prior experiences and knowledge
- capture children's interest
- be interdisciplinary, so that children see that reading, writing, mathematics, and other curricular areas are part of science and technology
- integrate several science disciplines
- serve as vehicles for teaching major organizing concepts, attitudes, and skills
- utilize the following instructional characteristics appropriate for kindergarten through eighth-grade science:
  Each is a hands-on activity, and students' lives and their world are the focus of the activities.
  Scientific and technological concepts and skills are developed within personal and social contexts.
  The students construct their concepts, attitudes, and skills through a variety of experiences.

This integrated approach is also consistent with the new National Science Education Standards, reached by a consensus of hundreds of teachers, scientists, and policymakers. These standards call for developmentally appropriate curriculum that is connected to other school subjects, coordinated with mathematics programs, and interesting and relevant to children (National Committee on Science Education Standards and Assessment, 1994).

Secondary Integrating Activities

Two important secondary methods of integrating learnings are less direct than the primary methods, since they depend upon a teacher's ability to plan ahead to apply them. The methods are (1) maintaining concepts by applying them, and (2) helping children connect new concepts to already acquired concepts. If the topics to be linked are presented sequentially whenever possible, children can be greatly helped to pull together relationships by themselves. The understandings gleaned from one set of activities can then lead directly or transfer indirectly to those that follow. As topics are developed in this text, some general suggestions are made for continuing these conceptual concerns and accommodating new information to already established concepts. Susan Carey's (1988) research on science education emphasizes that, "to understand something, one must integrate it with already existing knowledge schemata." These secondary integrations surround the primary integration model as depicted in Figure 1–2.

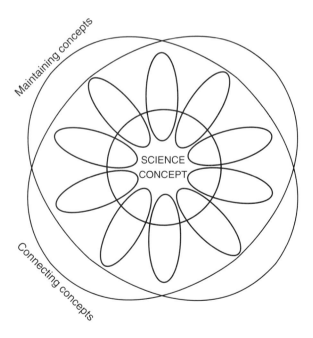

FIGURE 1–2

REFERENCES

AINSWORTH, M., & BELL, S. (1970). Attachment, exploration, and separation illustrated by the behavior of one-year-olds in a strange situation. *Child Development, 41,* 49–67.

BIBER, B. (1979). Thinking and feeling. *Young Children, 35,* 4–16.

BUZAN, T. (1994). *The mind map book.* New York: Dutton.

BYBEE, R., BUCHWALD, C., CRISSMAN, S., HEIL, D., KUERBIS, P., MATSUMOTO, C., & McINERNEY, J. (1989). *Science and technology education for the elementary years: Frameworks for curriculum and instruction.* Washington, DC: National Center for Improving Science Education.

CAREY, S. (1988). Cognitive science and science education. *American Psychologist, 41,* 1123–1130.

CARSON, R. (1960). *A sense of wonder.* New York: Harper & Row.

EPSTEIN, S. (1994). Integration of the cognitive and the psychodynamic unconscious. *American Psychologist, 49,* 709–724.

FRAIBERG, S. (1950). *The magic years.* New York: Norton.

GARDNER, H. (1993). *Frames of mind* (2nd ed.). New York: Basic Books.

HARLOW, H. (1958). The nature of love. *American Psychologist, 13,* 673–685.

NATIONAL COMMITTEE ON SCIENCE EDUCATION STANDARDS AND ASSESSMENT. (1994). *National science education standards: Discussion summary.* Washington, DC: National Research Council.

RAZRAN, G. (1961). The observable unconscious and the inferable conscious in current Soviet psychophysiology: Interoceptive conditioning, semantic conditioning and the orienting reflex. *Psychological Review, 68,* 126–127.

RESTAK, R. (1991). *The brain has a mind of its own.* New York: Crown Publishers.

SALK, J. (1994) *Bottom line* (Vol. 14, 4). Personal interview.

SELIGMAN, M. (1992). *Helplessness.* San Francisco: Freeman.

SEVERIDE, R., & PIZZINI, E. (1984). What research says: The role of play in science. *Science and Children, 21,* 58–61.

SPERRY, R. W. (1974). Lateral specialization in the surgically separated hemispheres. In F. O. Schmitt & F. G. Worden (Eds.), *The neurosciences: Third study program.* Cambridge, MA: M.I.T. Press.

Science Participants: Children, Teachers, and Families

YOUNG CHILDREN AS THINKERS

It is tempting to think of young children's minds as fresh pages to be written on, or clay to be molded by skillful, caring teachers. It is so tempting, in fact, that many inexperienced teachers and parents equate "telling" with teaching. Unfortunately, much writing about education characterizes teaching as "delivery of instruction," as if children's minds were loading docks upon which teachers can deposit boxes of information and skills.

A more realistic view of how children's minds work comes from an ever-widening stream of cognitive psychology research. This view is reinforced by the once-again-appreciated observations of parents and teachers who work daily with children. Both sources see children's minds constantly engaging in sense-making. Consider this story a mother reported to her son's teacher:

Three-year-old Christopher was baking gingerbread with his mother when he asked, "Where is the cinnamon-god?" His surprised mother probed gently to puzzle out the meaning of his question. She learned that the rabbi from a neighboring synagogue had recently visited the children at Christopher's preschool. God, synagogue, cinnamon, cinnamon-god: Christopher had put all the pieces together. He had constructed his own knowledge.

Most likely, Christopher had used previous experiences with cinnamon—something he could touch, smell, taste, and see—as a base of understanding on which to hook the new information. His mother then helped him refashion the connection to conform to more commonly held definitions and pronunciations of *cinnamon* and *synagogue*. Because his mother offered more information, Christopher could form new, more sophisticated knowledge. He could do this without being aware of how marvelously his mind was working. Because at his age he is learning effortlessly about nine new words a day, every experience Christopher has is a source of developing knowledge for him.

Cognition Theories

The role of active experience in learning was both theorized about and demonstrated by John Dewey and Maria Montessori early in the 20th century. Jean Piaget added to their work by focusing on details of the sequencing and content of children's thought development. His explorations led him and his followers to what is today's dominant cognition theory, *constructivism*. This is the view that children build knowledge internally by interacting with the world to learn how it works, and to make meaning of it. Piaget especially valued the development of logical-mathematical knowledge which is also of particular interest to science educators, since this type of knowledge is identified with science knowledge.

Piaget contributed immeasurably to the development of the early childhood education field by maintaining that young children think differently, in some circumstances, than do older children and adults. He believed that young children require a special kind of curriculum, because their thinking is more concrete and less logical. (One of the main reasons the very young think differently is their lack of experience with objects, events, people, and places.) Constructivism implies that there must be experiences from which to construct knowledge. Young children especially need these hands-on underpinnings of thought.

While Piaget postulated invariant, age-related stages of intellectual development, later research has indicated that young children are not as illogical and concrete in their thinking as Piaget claimed, nor are adults free of these intellectual dispositions (Carey, 1986). Rather, it appears that both children and adults—without sufficient experiences, education, and expertise—misconceive aspects of the physical and social worlds.

Not that we should be surprised at adult misconceptions. Science educator David Hawkins (1985) thinks that when it comes to science, most of us are in the 17th century, perceiving the world intuitively and commonsensically, despite the greater insights of more modern science. Howard Gardner (1991) adds that typically by the age of 5 years, we have constructed enough knowledge of the physical world to function in it capably, and much of this knowledge is very resistant to change by schooling.

It seems, then, that every person has constructed, and continues to construct, knowledge. Since some constructions are resistant to change, particularly in the physical sciences, the role of teachers is crucial. In what way do teachers matter? The classic, but newly influential, work of the Soviet psychologist, Lev Vygotsky, addresses this. It adds to Piaget's theories the insight that children are helped and influenced in their knowledge construction by the people around them. For instance, our preschooler, Christopher, will learn to distinguish between cinnamon and a synagogue because his mother shared her society's understanding that these things are distinguishable. She interacted with Christopher in ways that will bring him into that shared understanding.

Teachers, like parents, act as bridges between what society understands to be true and valuable, and what children are figuring out about their environment. Much of what children come to understand about their environment can properly be called *science;* gravity, evaporation, and changes caused by heat and cold, are examples. Helping children sort out their ideas, giving them access to common vocabulary related to the topic, and extending their thinking is the role of persons who know a bit more, such as teachers. Although many teachers feel uneasy in their understanding of science concepts, many of these concepts are amenable to joint discovery by teacher and children. Working with real materials, reinforcing the concepts in many different ways with different media, and being open to surprises all contribute to concept development in everyone participating.

For decades, early childhood educators have valued the practice of providing plenty of time and a wide variety of media to teach concepts. Recently this practice has received theoretical support from Howard Gardner's multiple intelligences theory. Gardner (1993) believes that intelligence is more than the single, logical-mathematical processing of stored facts that intelligence tests assess. He sees intelligence as problem-solving, problem-creating, and problem-finding across a range of situations. He identifies seven interlocking *kinds* of intelligence: logical-mathematical, linguistic, musical, spatial, bodily-kinesthetic, interpersonal, and intrapersonal. While it is not clear exactly how these intelligences develop independently and interactively, indentifying them has provided early childhood educators with a stronger rationale for using integrated curriculum. Those multiple approaches help children effectively construct and use socially valued knowledge, such as science.

In this text, we try to guide teachers in helping children construct powerful science understandings. By appealing to children's emotional and intellectual interest, by providing multidisciplinary approaches to concepts, and by modeling how teachers can spark and maintain children's engagement with science, we hope to encourage a community of learners. In this community, children and teachers work together to build understanding from their daily efforts.

YOUNG CHILDREN AS PERSONS

Earlier we looked at the influence children's feelings about themselves and their physical world have on their ability to think and learn. Learning, or failing to learn, is also strongly influenced by children's feelings about their place in the social world. The characteristic ways children feel about themselves and relate to others are central parts of their personalities. The classic work of Erik Erikson (1977) identified general developmental trends in personality growth that both contribute to and are enhanced by successful learning. These trends are significant affective components of cognitive development.

Sense of Initiative

If children's earliest experiences have promoted trust in others, and if children have then developed a good sense of autonomy, the next positive personality trend, *a sense of initiative,* unfolds. Three- and 4-year-olds who are developing a sense of initiative have great energy to invest in activity. They want to know what they can do. They are eager for new experiences and new information about their world. Emerging reasoning powers blend with a developing imagination, leading the child to ask searching questions. Children's delightful curiosity in this phase can be, and endlessly is expressed as "Why?" A budding behavior control system allows children to slow activity long enough to become immersed in events that capture their attention.

Adam is immersed in quiet insect observation.

Preschoolers seek the acceptance of children who are like themselves in some way and who have mutual interests. This sociability means that a preschool child rarely works alone at the science table, although sharing equipment and materials with others may still be somewhat difficult for them. In general, personality developments in this stage make it an ideal time to explore with children the regularities, relationships, and wonders of the world close at hand.

Sense of Industry

Young school-age children who have succeeded in developing a sense of initiative move toward the next positive personality trend, described by Erikson as the *sense of industry.* Eagerness to do meaningful tasks and to become good at them is evident. Children may persist in working at a task until it is completed, just for the sake of satisfaction in accomplishment. During this phase, children's sense of competence is enhanced by mastering challenges. These children are able to work cooperatively with others because they have good inner control over their behavior. The desire to follow the classroom rules is especially visible in kindergartners. If all goes reasonably well in children's lives, this positive personality trend continues through the remaining early childhood years.

Respecting Personal Development Traits

The natural tendencies to learn where one belongs and what one can do in the social and physical worlds, to create and imagine, to seek answers, to become increasingly effective as an active doer and thinker, and to take pride in independent accomplishment and growing competence all support the interest and persistence needed to learn throughout life. These tendencies and abilities, however, can be misdirected and even diminished to the point of burnout when they are channeled too early into formal training in academic skills and fact memorization (Elkind, 1987). These tendencies are respected and nurtured when preschools and kindergartens allow children to choose whether or not to participate in science explorations.

THE TEACHERS

Who Can Teach Science?

Any teacher who is capable of maintaining a classroom atmosphere of warmth, acceptance, and nurturance meets the basic qualifications for guiding young children in discovery science. In addition, positive attitude toward science and the ability to carry out the catalyst, consultant, and facilitator roles are needed for good teaching. However, little can be taught when a close rapport between child and teacher is missing. Children learn most from people with whom they feel the bond of personal interest and caring.

Attitude Contagion

Children's long-term attitudes toward science as subject matter begin with the attitudes of teachers whom children encounter in their earliest exposures to science. A teacher's positive attitudes toward science may have a long history. For some teachers, these good feelings accumulated during their own satisfying exposures to science in elementary and secondary schooling situations. Other teachers may have retained strong science interests in spite of inadequate school experiences. They may have had an "answer person" in their lives: a parent, a grandparent, or perhaps a camp naturalist who had interesting information to share and the patience to help a child find answers. People with such fortunate backgrounds feel comfortable teaching science, and they recognize the importance of science in their students' lives.

On the other hand, older children's negative attitudes toward science may have been "caught" from unenthusiastic teachers who relied on textbooks as the only source for gaining information. Those teachers may have been too unsure of themselves to allow children to consider alternatives to the textbook explanations, or to discover science concepts through experimentation. Many female students lost interest in science when they came up against gender bias in the classroom. They observed male students receiving preferential teacher attention because of the stereotyped assumption that males have a stronger aptitude for science (Sprung, Froschl, & Campbell, 1985). Still others have been discouraged by science courses that emphasized memorizing facts rather than understanding the principles that affect our daily lives.

Negative attitudes toward teaching science can be turned around by opportunities to observe and participate in hands-on science experiences with eager young children. Participation in activity- based science education courses or workshops can restore lost confidence and revitalize faded interest. Classroom follow-up studies of teachers who have participated in activity-based science workshops reflect shifts toward more positive attitudes. The majority of the 400 teachers who were observed as they taught were found to have greatly improved their skills and to have increased the amount of class time they devoted to science activities (Bredderman, 1982).

Authentic Interest

The teacher's authentic interest in finding out more about something is a vital part of the positive teaching attitude. This interest implies a willingness to learn along with the children when they lack answers. The ability to admit to not knowing everything was one of the traits of a good teacher, as identified by older students in a cross-cultural study (Hofstein, 1986).

The teacher's own interest in finding out sustains the children's ready curiosity. It can revive curiosity in children who have been belittled for asking questions, and it can rebuild curiosity that has atrophied in an unstimulating environment. When the teacher's own sense of wonder is alive and active, curiosity behavior is modeled for the children. This important attitude has a fundamental place in the science-teaching framework, as indicated in Figure 2–1. If the teacher is indifferent about how things work, why should the children care?

Teaching Roles

Effective discovery-science guidance calls for four teaching roles:

1. The *facilitator* creates a learning environment in which each child has a chance to grow. Planning, gathering needed cast-off materials, and actually trying experiments are science facilitator tasks. In this role there is a tolerance for messiness as children work, a willingness to risk new ventures, and an ability to profit from mistakes.

2. The *catalyst* turns on children's intellectual power by helping them become aware of themselves as thinkers and problem-solvers. This role contrasts with the "teacher" image so many of us carry from our own school days in which the teacher seemed to be the ultimate source of all knowledge. A teacher like this

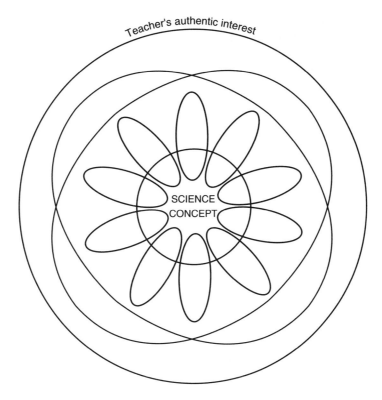

FIGURE 2–1

can dim children's intellectual power by somehow magnifying the distance between the knowledge in his or her possession and that of the child. The catalyst, on the other hand, sets a positive, encouraging tone by staying in touch with his or her own excitement in discovery.

3. The *consultant* observes carefully, listens closely, and answers questions simply while children engage in their explorations. In this role, small bits of information can be offered as learning cues, and questions can be asked of the child to help him focus on the relevant parts of a problem. The consultant allows the child time to reflect on the new idea and tackle the solution independently. This role often intimidates beginning teachers until they can accept themselves as learners, too. The consultant role is a supportive coaching role rather than a directive one.

4. The *model* deliberately demonstrates to children the important traits of successful learners such as curiosity, appreciation, persistence, and creativity. Lillian Katz (1985) defines these characteristics as *dispositions:* habits of mind or tendencies to respond to situations in particular ways. Katz points out that, although

these qualities are essential, adults rarely identify and demonstrate these dispositions toward learning; yet, these dispositions are best learned through example. We model these positive learning dispositions when we share our personal experiences and thought processes with children. We model persistence by describing our own efforts; for example: "At first I couldn't make it work, but I just kept on trying until it did work." We model creativity when we reveal a problem we solved in a new way; for example: "I needed a better way to let you see both sides of the moth. Then, I had an idea that worked. I taped these two plastic lids together to make this neat display case for the moth."

The teacher who holds positive attitudes and understands his or her necessary roles is well-rewarded by both personal growth and by the pleasure of sharing the sense of mastery that children find through their science discoveries.

THE FAMILIES

All parents have information about the world that can be passed on to their children—to their mutual benefit. Jim's dad let Jim scrub tires and wipe grease off bolts for him as he salvaged parts from the derelict cars that stood in their front yard. As a consequence, Jim was able to make contributions to class discussions about simple machines after weeks of virtually silent presence in school. Daniella's grandmother contributed the cecropia cocoon that, months later, provided us an hour of breathless wonder as the moth struggled free in the full view of 20 children. Five-year-old Tim was able to share solid information about many kinds of planes and how they flew, because his pilot father took him to hangars to see the planes and watch the mechanics at work. These family members enriched the learning of a whole class. They also intensified the interest of their own children in knowing the world because they endorsed the value of that knowledge.

The teaching role of the parent (or parent figure) is unique and vital. Parents, not classroom teachers, have the only continuous opportunity to guide a child's intellectual growth. The earliest imprinting of attitudes of appreciation or disregard for the natural or man-made world takes place within that primary relationship. This teaching by example occurs as part of the child's socialization, whether consciously or not.

The power of parents' supportive interest as motivation for children's immediate learning achievement has long been documented (Crandall, 1965). More recent evidence demonstrates the long-lasting effects of parental encouragement on the degree of interest of women and minorities in college math and science (Thomas, 1986). Bloom's 1985 study of the early development of successful, talented adults (including scientists) consistently revealed parental support for the child's special interest, as well as encouragement to persevere and pursue those interests wholeheartedly. Our own science teaching can help generate parental interest where it might be lacking. When children take home evidence of practical problem-solving ability or share science information that has meaning in the adult world, parents rarely fail to respond positively.

Since its inception, early childhood education has striven for a teaching partnership with parents. In response to the National Education Goals for the year 2000, the U.S. Department of Education has focused on enlarged parent involvement as one way to revitalize science education. It recently brought out a booklet for parents, *Helping Your Child Learn Science,* to promote this goal. (See Appendix B.) The National Science Teachers' Association (1994) adopted a position statement on parental involvement in science education. It gives priority to encouraging parents to help their children by seeing science everywhere, doing science activities together, and taking advantage of community resources of many kinds, efforts this textbook has been advocating since 1976, incidentally.

Unfortunately, many parents do not realize that they possess science knowledge that could be comfortably shared with their children. They may never have experienced the companionable pleasure of watching an impressive caterpillar with their child. They may not have paused to take a close look with their child at the fascinating gears on a bank vault door. To help parents promote children's interest in science, informal, enjoyable Exploring at Home activities have been described for each of the major science topics presented in Part 2 of this book. They are found in Appendix B.

These projects are intentionally presented in a lighthearted way to avoid inappropriate performance demands by parents that discourage rather than encourage their children's efforts. Each of these related activities may be duplicated by classroom teachers and sent home with the clear intent of offering family fun, rather than required homework. To emphasize the special role the family plays in

Mother lifts Maya for a closer look at dozens of gleaming gears. She validates and rewards Maya's curiosity.

children's education, most of these activities are those that are best done at home at night, that directly relate to the home, or that are long-term and encompass beginning-to-end cycles that can have special value for families (Furman, 1990). A bibliography listing home-based science project books of special interest to parents appears in Appendix B. The influence of parental support completes the science learning framework, as shown in Figure 2–2.

A growing number of cities now have children's museums featuring interactive, physical, and natural science discovery areas. Many are specifically designed for family involvement. Others, like the Fort Worth Museum of Science and History, include a full schedule of informal classes from preschool through high school. Teachers can inform parents about the nearest science museum or nature center that features exhibits for children.

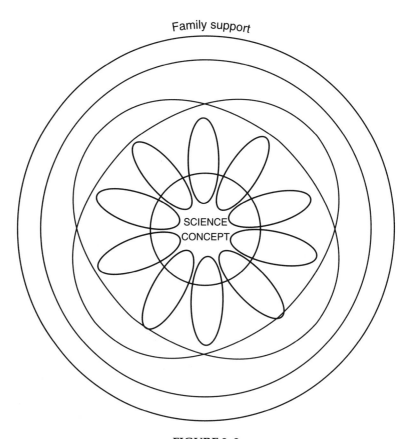

FIGURE 2–2

REFERENCES

BLOOM, B. W. S. (Ed.). (1985). *Developing talent in young people.* New York: Ballantine Books.

BREDDERMAN, T. (1982). What research says: Activity science—The evidence shows it matters. *Science and Children, 20,* 39–41.

CAREY, S. (1986). Cognitive science and science education. *American Psychologist, 41,* 1059–1063.

CRANDALL, V. (1965). Achievement behavior in young children. In W. W. Hartup & N. L. Smothergill (Eds.), *The young child: Reviews of research* (Vol. 2, pp. 165–185). Washington, DC: National Association for the Education of Young Children.

ELKIND, D. (1987). Superbaby syndrome can lead to elementary school burnout. *Young Children, 42,* 14.

ERIKSON, E. (1977). *Childhood and society.* New York: Norton.

FURMAN, E. (1990). Plant a potato—Learn about life (and death). *Young Children, 46,* 15–20.

GARDNER, H. (1993). *Frames of mind* (2nd ed.). New York: Basic Books.

HAWKINS, D. (1982). *Conceptual barriers encountered in teaching science to adults: An outline of theory and a summary of some supporting evidence.* Report to the National Science Foundation, Washington, DC.

HOFSTEIN, A., SCHERTZ, Z., & YEAGER, R. (1986). What students say about science teaching, science teachers, and science classes in Israel and the U.S. *Science Education, 70,* 21–30.

KATZ, L. (1985). *Dispositions in early childhood education.* ERIC/EECE Bulletin, 18.

NATIONAL SCIENCE TEACHERS ASSOCIATION. (August, 1994). *NSTA position statement: Parent involvement in science education.* Washington, DC: Author.

SPRUNG, B., FROSCHL, M., & CAMPBELL, P. (1985). *What will happen if . . . young children and the scientific method.* New York: Educational Equity Concepts.

THOMAS, G. (1986). Cultivating the interest of women and minorities in high school mathematics and science. *Science Education, 70,* 31–34.

VYGOTSKY, L. S. (1962). *Thought and language,* New York: Wiley.

3

Guiding Discovery Science

DISCOVERY SCIENCE

In discovery science, learners actively seek information. Regardless of the level of sophistication, be it science in the research laboratory or in the preschool, science is a way of thinking and gaining knowledge that consists of four related steps:

1. Noticing (becoming aware of a problem)
2. Wondering why (hypothesizing, proposing an explanation)
3. Finding out (experimenting)
4. Sharing the results with others

Science activity can also be described in terms of the methods or processes used to carry out scientific activity: observing, classifying, measuring, communicating, experimenting, predicting, inferring, designing investigations, interpreting, and constructing explanations from the data. The least complex of these processes—observing and rudimentary experimenting—can be seen in the behavior of infants. As their thinking develops toward the final stages of abstract thought, children can use these processes with increasing sophistication.

Some science educators advocate basing primary and elementary science programs entirely on the development of skill in using these processes, without focusing on specific science content. Others feel that to separate the learning process from that which can be learned denies children the opportunity to satisfy curiosity and to acquire helpful information. It is quite possible for children to become proficient at using the science processes, while at the same time gaining concepts through hands-on discovery activities.

One of the best means of helping children know the world at hand is to organize materials so children can explore, question, reason, and discover answers through their own physical and mental activity. This guided discovery approach to science learning emphasizes *how* to find answers, as well as *what* can be learned. When discovery science experiences are seen as part of the child's continuous search for knowledge, it makes good sense to support and enable that search in the classroom in these basic ways:

- Discovery science values and rewards curiosity as a valid learning tool (Kyle, Bonnstetter, McCloskey, & Fults, 1985).
- Discovery science encourages individuality and creativity in children's problem-solving. This leads to better retention of the concepts attained. Interestingly, the four steps of science as a way of knowing and thinking are the same ones used to describe the creative thinking process.
- Discovery science validates the fundamental learning style of direct involvement with materials. It builds upon spontaneous experiencing, touching, and trying by the very young. Active participation remains so important to knowing throughout life that our language uses the term *firsthand* to mean something that is known from the original source.
- Discovery science experiences in early schooling help to gradually replace the young child's intuitive explanations of the unknown. Such experiences demystify bewildering events, but respectfully retain the profound and beautiful aspects of natural occurrences.
- Discovery science provides a means of focusing the attention of the restless child, the anxious child who is preoccupied with personal concerns, and the bored child who is understimulated by less challenging aspects of the curriculum.
- Discovery science has appeal for the resistant learner who has come to protect herself from inappropriate pressure to perform formal learning tasks. The play-like quality of exploring real materials reduces the stress that more precise paperwork tasks produce. Discovery science allows more physical and social involvement than do the more structured forms of schoolwork. Pressured children can put aside their resistance and restore their sense of themselves as learners when they engage in science activities.
- Discovery science, by its very nature, provides an intriguing path to the goal of developing children's intellectual potential. Any form of learning that includes the manipulation of interesting materials is highly motivating to children. The discovery science activities and cross-curricular extensions in this book lend themselves to programs where language learning and content are integrated, such as the language experience approach, the whole-language approach to reading, and the project approach to curriculum development. The sensory and psychomotor extensions fit well into ESL (English as a second language) programs that emphasize total physical response as a learning strategy.

GUIDING LEARNERS

Teaching Styles

Opinions differ about the teacher's role in experiential learning. Some educators interpret the Piagetian maxim, telling is not teaching, quite literally. They consider that any verbal instruction from a teacher interferes with the child's genuine understanding of new ideas.

Other educators and researchers assign a more active role to teachers in the child's acquisition of knowledge. They believe that children do benefit from appropriate forms of verbal guidance. Robbie Case (1986) points out that a teacher who can empathize with a child's efforts and share a child's feeling of excitement and satisfaction in solving problems is better able to guide and encourage children's intellectual development. Case believes that children learn by imitating and interacting with others, as well as through their own exploration and problem-solving. He sees the teaching role as social facilitation of children's problem-solving abilities.

Guided discovery learning calls for this social facilitation as direct and indirect guidance. The teacher supplements the child's active exploration of problems by helping the child draw meaning from the experience and by extending conceptual learning. The crucial affective element here is the teacher's belief in the importance of this effort. Without commitment to a central purpose for teaching science, our lesson plans fail to inspire children; our hands-on activities lose vitality. When we are committed to the spirit of sharing with children the powerful constants that make their world more predictable, science comes alive in the classroom. When our focus shifts from teaching facts to teaching children, the process of helping to uncover enlightening knowledge remains fresh and dynamic.

Organizing Approaches

There are two distinctive approaches to organizing socially facilitated science activities: child-instigated experiences (the incidental approach) and teacher-instigated experiences. The child-instigated approach is ideal for very small classes of preschool children, and for nurturing the individual interests of gifted older children.

Incidental science can occur at any time or place, whenever a child's curiosity is aroused by something significant: a sparkling bit of quartz embedded in the sidewalk, the iridescent sheen of a beetle's back, or an evaporating playground puddle. The teacher capitalizes on the child's discovery by asking questions that lead to further discovery, by relating the find to something the child already knows, by extending the experience into other classroom activities, and by offering to help the child locate other resources for expanding his or her information. The confident teacher with a ready interest and strong background in science can do an admirable job of enlarging children's information this way.

There are drawbacks to this approach, however. It is difficult to provide the same degree of support for larger classes, or to fit spur-of-the-moment activities into a more structured school program. Nor do these incidental, child-instigated investigations constitute an adequate conception of science. Science is not only a collection of facts about the physical world, but also a way of describing and explaining physical phenomena and organizing them according to relationships among them.

The teacher-instigated approach outlined in this book centers on interests common to young children. It places related experiences within a framework of corresponding extensions. The science experiences themselves are not intended for random presentation as interesting time-fillers. Rather, the experiences and suggested extensions are presented in a unified way, perhaps continuing for a week or two. It is tempting to regard units of learning as accomplished work that can be neatly tied up and shelved for the year. However, lasting learning takes place through many encounters with the same ideas in a variety of contexts. Science learnings can be referred to and renewed throughout the school year when they relate to other learnings and classroom occurrences. Recent research has demonstrated that material reviewed and built onto over a period of time can be retained remarkably longer than can isolated fragments of material that have only been presented once (Adler, 1991). Science knowledge further takes hold when it has utility and when it becomes part of a continuous chain of learning that develops throughout life.

PREPARATION FOR TEACHING EARLY SCIENCE

There is no question among educators about the superiority of active discovery over the textbook and lecture method of learning science. Numerous comparative performance studies conducted over the past 25 years have confirmed that superiority; yet it appears that most kindergarten and primary grade teachers continue to use the less effective methods. The most common reason teachers give for not offering hands-on learning activities is their conviction that they lack enough science background to be able to answer children's questions. The open-endedness of problem-solving bothers them. The feel uneasy when more than one solution can be appropriate. When these science anxieties go unchecked, they can be easily transformed into a list of airtight reasons for reading about, instead of "doing," science.

A sense of inadequacy about science can be a real limitation to science teaching because it tends to be communicated nonverbally to children as a distaste for the subject matter. However, it is equally possible to do a poor job of teaching science concepts to youngsters because of being over-trained in science! It can be difficult for sophisticated scientists to translate their verbally elaborate concepts to a level of ideas that young children can understand.

Actually, most of us have much more practical science information than we may recognize. We can fill in knowledge gaps as we prepare to teach. A beginning teacher can supplement basic knowledge by carefully reading the *concepts, learning objective statements,* and *activity directions* in this book. Trying out the experiments and recording the results, problems, and personal reactions to the experience before presenting them to children can increase a novice's confidence. Reading some of the starred reference books, written at the young child's level of understanding, will add appropriately to one's knowledge base. Reference books suggested for teachers will provide depth of information to enlarge the teacher's developing interests and to provide answers prompted by a child's urgent curiosity.

The recommended children's reference books and science-based stories for children—especially those by Aliki, Franklyn Branley, Joanne Ryder, and Rose Wyler—are written by authors who know both the science concepts and the young learner's capacity to absorb core ideas. Many of these fine writers also know how to incorporate humor and excitement into their work. Other resources were chosen primarily to supplement material in this book with additional activities that children can accomplish easily. Some were included as sources of hard-to-find illustrations of science principles at work, or to provide challenges for children who are fascinated by a particular topic.

There is merit in renewing and strengthening our science information just before we teach. Then the satisfaction gained in becoming reacquainted with the concepts will be fresh and available to spark the interests of children. Finally, it is reassuring to remember these words by Jerome Bruner (1961): "The basic ideas that lie at the heart of all science are as *simple* as they are *powerful*" (p. 12).

INDIRECT AND DIRECT TEACHING

Discovery learning is guided *indirectly* through thoughtful questioning and listening, and by sensitive discussion-leading. It is guided *directly* by offering conceptual cues and by encouraging effort. Since many of us were not exposed to these kinds of guidelines in our own schooling, it will be helpful to analyze these techniques.

Thoughtful listening to children's ideas is an indirect way to guide and sustain their interest in discovery learning. We do this when we avoid passing judgment on a child's erroneous ideas or private logic that differs from a scientific explanation. Susan Carey's research (1986) confirmed that preschool children can have already formed naïve, intuitive explanations of biology based upon their own interpretation of experiences. There are ample anecdotes from teachers and parents to suggest that the same is also true of other topics that intrigue children. When there are gaps in the information they can assemble, children typically fill in those gaps imaginatively to form their personal theories, as did Matthew whom we met in Chapter 1. This process is not unique to children, however. Countless adults still naïvely believe, from the evidence of their own perceptions, that the sun orbits the Earth. Each of us may be able to recall some of our own "childish" misperceptions, ideas that we perhaps dared not express for fear of being laughed at. Had we felt it was safe to reveal them, those misperceptions might have been cleared up much sooner. Children's confidence is sustained when we respect their naïve attempts to figure things out instead of devaluing them as amusing or quaint, or as serious misconceptions that must be eradicated immediately. Intuitive theories can be gradually clarified and reorganized when discovery experiences reveal the more systematic and complete way of looking at things that we call science.

We indirectly guide children's thinking when we show them our own reasoning processes by "thinking out loud." That could occur when we say something like this: "When I found this tiny leaf on the sidewalk, I thought to myself, 'This isn't like any other tiny leaf I've seen before—Hmmm. But it has five points, like a star. A big sweet gum tree leaf has that shape, too. I wonder if a new sweet gum leaf starts out as tiny as this.' So, what do you think I did to find out for sure?" Sylvia Farnham-Diggory (1990) describes this important modeling as making invisible thinking processes visible to children.

LEARNING TO QUESTION

Just as the quality of scientific research depends upon asking worthwhile questions, the quality of discovery science learning is indirectly guided by helpful questioning. Open-ended questions are useful for generating several appropriate answers (divergence) and for engaging both left- and right-brain processes. *Divergent questions* can serve many purposes:

- *Instigating discovery:*   A science activity becomes a discovery challenge when it is initiated as a question to answer. Each activity suggested in the following chapters is headed in bold type by the question the activity can answer. These questions can be used both to introduce the activity and to elicit children's conclusions about their activity. The questions can be printed on index cards and posted in the science area to help reading children stay focused in their work.
- *Eliciting predictions:*   Before children experiment, elicit their predictions: "What do you think will happen if. . . ?"
- *Probing for understanding:*   "Why do you think that side of the balance went down?"
- *Promoting reasoning:*   "Why do you think this arm feels dry and this arm feels wet?"
- *Serving as a catalyst:*   Sometimes a question can be a catalyst that sparks renewed interest in a problem, wheareas a specific direction from the teacher could make finding out unnecessary. "What could you change to try to make your lever work?" encourages new effort. "If you move the block closer to Robert, you'll be able to lift him with your lever" discourages independent effort.
- *Encouraging creative thinking:*   In a group discussion ask questions such as, "How would our lives be different if there weren't any friction?"
- *Reflecting on feelings:*   In a group discussion, ask questions such as, "What was the best part of the dark box experiment?"

There is ample research demonstrating that teachers tend to overuse *convergent questions,* that is, those with a single, correct answer. Convergent questions tend to draw primarily from left-brain processes and are less effective in stimulating creative thinking or synthesizing; however, it can be useful and suitable to ask certain convergent questions to promote learning. Convergent questions are beneficial for

- *Directing attention:*   By asking a question like, "Does the red cup hold as much water as the blue cup?" we can direct a child's attention to a key part of the activity she has overlooked. The child can then correct her own course of action without feeling criticized.

- *Recalling the temporal order:* "What did you do first? What did you do next?" "When did this happen? What happened afterward?"
- *Recalling prior conditions:* "Does the jar of beans look just the same today as it did yesterday?"

Follow-up questions to convergent questions can be divergent. They can lead to further reflection or to fresh experimentation: "How can you find out?"

It is helpful for beginning questioners to role-play with other beginners to improve and feel at ease with new techniques. To overcome habitual closed questioning, it can take concerted effort to change the way we shape our questions. Many of us use the pattern of stating in the form of a question the answers we wish to hear from children. We may say, "So the cup of snow melted into a smaller amount of water, right? Isn't that what you found?" This style of questioning reduces children's need to discover answers for themselves, or tells them that the main discovery is to find out what the teacher wants them to say.

LEADING DISCUSSIONS

Discussions serve different purposes at different points in learning new concepts.

- *Introductory discussions* whet interest in a new topic when children are encouraged to recall events they have encountered personally and to contribute what they already know about the subject. At this time children are encouraged to raise their own questions about what they would like to find out in the activities.
- *Small-group discussions* follow the children's individual activity to process what took place in their experimentation and to help clarify their thinking. These discussions can reveal that more than one conclusion can be reached from an activity.
- *Summary discussions* with the whole class pull together the concepts that have been explored in the activities and extensions. Each of the following chapters offers suggestions appropriate to summary discussions, under the heading *Secondary Integrations.*

Group discussions allow children to learn from one another, if the teacher models respect for the ideas and experiences that children express. To encourage interactive responses among children, quiet children may have to be drawn out in low-risk ways. Recognize a nod of agreement or a responsive smile from such a child as involvement: "It looks like Danny agrees with your idea." If some children have trouble staying involved in large-group discussions, this concrete illustration can be helpful:

> Ask two children to each get a crayon and then exchange the crayons with each other. Have them verify that each had one crayon *before* sharing with the other, and each still has only one crayon *after* sharing with the other. Next, ask each child to share an idea with the other on a topic such as the day's weather. After each has shared an idea, point out that now each child has two ideas about the day's weather. Add, "We grow in ideas when we share and listen to each other."

It is important not to close off the exchange of ideas as soon as a child offers key information. Keep the discussion open until each willing child has had a chance to be heard, even if the ideas begin to echo one another.

Adults often overlook the fact that it takes a little time for children to produce thoughtful answers to questions. Many people feel they are unsuccessful at teaching if their questions are not answered immediately. The results of Mary Budd Rowe's (1974) seminal research indicate that just the opposite is true. She found that the teachers she observed gave children an average of under one second to answer questions. Teachers who were then trained to wait three seconds or more for responses elicited a greater number of answers, longer answers, and more varied and accurate responses. Furthermore, as the teachers waited longer, children began to listen to and respond to each other's comments. Children other than the "brightest" in the class also began to contribute answers. This suggests that when quick responses by a few children end a group discussion, many untapped ideas may be cut off. This practice leaves many children feeling less confident about themselves as thinkers and learners.

A good discussion puts the teacher in the catalyst role by empowering children's ability to think and express their ideas. This seems to be a difficult shift for some teachers to make, especially if they are accustomed to having their voices dominate the classroom. With practice, it becomes easier to support and elicit children's comments by using bridging remarks like, "That was a good idea, *and* others may have different thoughts about it. . . ."

The catalyst teacher underscores what children accurately contribute, adds bits of information to expand those ideas, and clarifies misconceptions that might still linger. The teacher summarizes the various points children make in the discussion. If the contributed ideas do not include certain salient points, the teacher can add, "Scientists also tell us that. . . ." Group discussions are most successful when they are guided with the goal of stimulating children's thinking and reasoning powers.

ORGANIZING TIME AND SPACE FOR SCIENCE

The logistics of making exploration time available to small work groups will vary for each classroom. Now that learning centers are more commonly used, teachers are becoming adept at flexible activity scheduling. Problems of managing science activities can be eased with the help of an assistant in the room, whether it is a volunteer parent, a retiree, or an "exchange student" from an upper grade. It is best to delegate overseeing ongoing activities to the helper. In some instances, simple directions typed on index cards or tape-recorded directions can allow children to handle science projects fairly independently.

A good location for science activities facilitates thinking by inviting children to participate and by controlling distractions. Proximity to storage and cleanup aids is important. Varying the setting for activities can build interest. Anticipation is heightened on days when science takes place under a blanket-covered table!

Elaborate bulletin boards aren't needed to attract children's attention to science when a frequently changed, PLEASE TOUCH display shelf is available. Many teachers begin the school year with noble intentions of welcoming nature finds and other objects of science interest that children bring from home. Perhaps they initiate the project attractively with a bird's nest propped in the crotch of a small tree branch, some special rocks, and a recently shed snake skin. If the goal of a changing display is forgotten, however, the old things will lose their meaning. Since there is little appeal in a dusty nest or a tattered snake skin, it is better to retire the too-familiar objects. An easy way to keep the science display pertinent to ongoing activities is to try to display indestructable materials being used in the discovery activities.

INTRODUCING SCIENCE ACTIVITIES

There are many creative ways to focus children's attention on a new science activity. A spur-of-the-moment thinking game could build interest: "There's something that we see in our room every day that we will use in science today. It has a handle on the outside and rollers on the inside." One relaxed teacher intrigues his students as he patiently searches through his pockets for the "something I thought you would like to know about." Opening a brown bag to reveal an unexplained object of science interest can provide enticement to find out more. Any of these attention hooks need to be followed with, "What do you know about this?"

A topic can be introduced by reading a related story or poem, and then following it with an open-ended question to be answered by the discovery activity. For example, the topic of light could be introduced by reading Russell Hoban's *Bedtime for Frances* and then asking, "Why do you think Frances was confused and scared by her robe on the chair?" A simple description of an occurrence could start the explorations. "Today, when I stood on my porch, a heavy, noisy repair truck rumbled by and I felt the porch shake. Have you ever felt shaking near something noisy?. . . Today you can find out more about shaking and sounds happening together." While it is not advisable to oversell science as fun and games, with a little care it is impossible to avoid making science seem an irrelevant chore.

GUIDING EXPLORATIONS

Hands-on for children should imply, as much as possible, *hands-off* for teachers. There will be occasions when a tactful offer to steady a screwdriver or to knot a parachute string can help a discouraged child achieve success, but unsolicited help should not be given in the interests of saving time. Butts and Hofman (1993) speak about the difficulty of "unlearning" established ideas. They urge teachers to listen carefully to how children are describing and interpreting their explorations. The authors point out that it is the conversation with children that occurs after a hands-on activity that makes a difference in their thinking.

Supportive teachers provide sensitive guidance to children's explorations.

The consultant role is a delicate one: quietly offering a reasoning cue to highlight the important part of an experiment, or asking a question and allowing the child time to reflect on the answer. (Reasoning cues can be taken from the concept statement headings for activities in this book: "You pushed *down* to lift him *up* with the lever.") Wise consultants resist the temptation to hurry a child along to discover what they, the consultants, already know. They are careful not to smother a child's tender spark of inquiry with a heavy blanket of directions and facts. Experiential learning is weakened when a child is rushed to a premature conclusion or "helped" out of arriving at the child's own solution.

SUPPORTING INTEREST IN KNOWING

It would be unrealistic to assume that the spontaneous desire to know will consistently lead all children in a group to take part in every science activity made available to them. It would certainly abuse the developmental goal of encouraging autonomous decision-making to require the whole group to engage in each science event that the teacher presents. Knowledge of the uniqueness of individual development should make it clear that each child brings different capacities and motivation into learning situations.

Subtle directives from families can strongly influence the choices children make in school. Direct or indirect messages like, "Be careful with those good clothes . . . Boys don't cook . . . Girls are afraid of spiders and worms," can be powerful inhibitors to children who are unsure of their affection rating with their parents. Such directives may curb a child's interest in trying new experiences.

Children also may have needs that take precedence over the desire to find out. Children who are lonely at home may be very reluctant to pass up any chance to play with special school friends. They may choose a science activity only when friends make that choice. Children who are tired, hungry, or feeling under par may not be able to summon enough interest to observe what is happening at the science table.

If a child consistently spurns science activities, it would be well to think about possible causes and consider ways to make participation easier. The teacher may simply need to ask for that child's help while organizing materials for the day's science activity. Children usually take a proprietary interest in a project that they have helped to prepare. However, when a science topic becomes a central theme threading through many parts of the school program, even children who do little active exploring will become aware of the concepts being presented.

Teachers positively influence children's interest and involvement with science when they value and reinforce the children's focus on learning. They do this when they use specific words of encouragement to commend children's efforts to observe and explore. This reinforces what a child is attempting and accomplishing. Sensitive teachers also do this by recognizing that a word of encouragement after a failure can have a more powerful effect on a child than does praise for success.

A child's involvement in a science activity will be enhanced by specific verbal support for what she is learning. Words of encouragement that validate children's competence promote a positive disposition toward science. They strengthen children's concepts of themselves as learners and contribute toward their interest and continued effort to find out more. When the encouraging comments focus on learning and the pleasure of accomplishment, children are likely to persist in exploring, even in the face of obstacles, because they are trying to understand something new and become more competent (Dweck, 1986).

There are some subtle, but significant, differences in the way praise is used and the effect it has upon learners. For example, words of encouragement are discouraging if they are spoken with hostility or sarcasm. True encouragement is wholehearted, with the tone of voice and inflection matching the meaning of the words. Also, many teachers are convinced that they are motivating children's interest in science by routinely issuing automatic praise. Instead, children may feel that they are being manipulated when praise is overdone.

Teachers who value and tend to reinforce correct performance by their classes may weaken children's learning motivation by using automatic, ambiguous praise to keep children "on task." Unfortunately, excessive, empty praise seems to interfere with children's spontaneous interest in an activity. Children's attention becomes directed toward proper performance to win the teacher's approval rather than toward what they are learning, and they lose interest in the activity for its own sake. In fact, such needless focus on how well children are working may interrupt the focused concentration needed to attain a satisfying outcome. Eventually such inappropriately motivated youngsters tend to give up on tasks when they encounter obstacles.

TABLE 3–1   Authentic vs. Automatic Encouragement

Situation	Automatic Praise	Authentic Encouragement
Amelia checks her seed sprouting jar. She exclaims, "My sprouts have leaves!"	"Terrific!"	"You really looked carefully to find the new growth."
Jake beams with joy when he gets two loads balanced on the scale.	"That's right!"	"It feels great to get both sides balanced."
Nina reports doing yesterday's light-bending experience a new way at home. It worked!	"Superscientist!"	"How exciting to find a new way to do it! Tells us more!

They may only keep working as long as the teacher is nearby to do the "motivating" for them. Extensive research on the fallacy of this approach is documented in Alfie Kohn's book, *Punished by Rewards* (1993). Differences between authentic and automatic encouragement are illustrated in Table 3–1.

Authentic encouragement addresses the positive feelings of satisfaction and accomplishment experienced by the child. When children invest their attention deeply in an activity and achieve a meaningful result, or when they go beyond what is expected of them and uncover something new, they grow. They feel the joy of meeting a challenge. These rewarding feelings of increased personal competence are the genuine sources of intrinsic learning motivation (Csikszentmihalyi, 1990).

DISPELLING STEREOTYPES

Gender and minority stereotypes about intellectual capabilities perpetuate self-doubt and create needless limitations to the development of this nation's problem-solvers. Teacher's actions and beliefs contribute heavily to the self-attitudes being formed by their students. Self-attitudes, whether negative or positive, are the single most crucial force in shaping what an individual is able to accomplish.

While we may voice our desire to help each child develop her or his intellectual potential, our actions may belie our words. Our actions are based on our values, some of which we hold unconsciously. We may be acting on buried, left-over prejudices about the thinking capacities and "natural" interests of women and minorities that were passed on to us as schoolchildren. Those stubborn inheritances need to be brought into our awareness and deliberately discarded before they can be replaced by reality-based information. We will truly be acting on our convictions about the problem-solving potential of each child we teach when:

It might by yucky, but it still fascinates the girls.

- Girls and minority children receive as much verbal and social encouragement from us as do Caucasian boys to answer science questions, contribute to science discussions, and persist at experimenting (Blake, 1993).
- The first chance to try new activities is not the exclusive right of Caucasian boys.
- Science professionals who visit our classes, or whose pictures are posted on our bulletin boards, are as likely to be female and non-Caucasian as they are white males.

When we communicate in these ways our belief that each child can participate effectively in some form of science experiences, we will be making a valid effort to combat crippling stereotypes (Blake, 1993).

ADAPTING EXPERIENCES FOR YOUNGER CHILDREN

While the activities in this text's framework were planned for children 4 through 8 years of age, many of the experiences can be adapted for younger preschool children. Two-year-olds can enjoy simple sensory explorations such as feeling air as they move it with paper fans, spinning pinwheels, or swinging streamers on a breezy day; feeling rock textures and weights; touching ice, then touching the water it melts into; watching, then moving like a goldfish; tasting raw fruits and vegetables that have grown from plants; listening to loud and soft sounds; or gazing through transparent color paddles to see surroundings in a new light.

Three-year-olds might be expected to engage in similar activities, taking in greater detail. They will be able to direct deeper attention to such things as exploring new dimensions with a magnifying glass. This group can enjoy some of the classifying experiences on a beginning level: sorting rocks from objects that are not rocks; things that float from those that do not float; and objects that are attracted by a magnet from objects that are not attracted.

ADAPTING EXPERIENCES FOR CHILDREN WITH SPECIAL NEEDS

Science is an approach to thinking and behaving that has value for children at any level of motor, behavioral, sensory, communication, or mental functioning. Because discovery science calls for collaboration among children, it fosters the personal interactions that children with special needs often miss. It can help classmates recognize the limited child as an individual with ideas and talents to contribute. Nobody laughed at Lori once her worm-scouting skills came to light. Her problems with pencils and numbers and letters didn't interfere with her passion for capturing insects. Her generosity in sharing them improved her social relationships. Scott's "slow" right leg and droopy right arm didn't limit his deep involvement in science. Frequently left behind on the playground, Scott was in his social and intellectual element at the science table. With good-humored inventiveness, Scott found his own way to water our plants. He soaked a clean sponge in the pitcher of water that was too heavy for him to manage. Then he squeezed the sponge for each plant, giving them a simultaneous drink and sponge bath!

Adapting materials to make them more manageable can enable success for a child with special needs. A plastic basting syringe is easier to use than a small medicine dropper for a child who finds it difficult to grasp objects. Bamboo toast tongs may help children with fine motor control problems to pick up small objects.

Changes in approach are important for children with vision and hearing impairments. Advance information about what is going to take place helps the visually impaired child deal with new experiences comfortably. Face-to-face communication is necessary with the hearing impaired child. Because a hearing loss is invisible to others, it is easy to forget that communications are cut off when that child's back is turned.

An easily distracted child benefits from a simplified, clear arrangement of materials to focus upon. Tangible boundaries, such as individual trays of equipment, can help the child with poor control feel assured of a fair share of things to work with. The teacher's close presence can help ensure an island of calm around these children to help them pursue their work. Tiny pebbles or seeds that could be inserted into ears or nostrils should not be available without supervision to children who lack safety awareness. Time given to planning for successful activities minimizes the time and attention such vulnerable children demand when frustrating tasks lead to disruptive behavior.

A chair-confined child will need to have materials brought into close range. A highlight event might be connecting a child with attention to invest and a grasshopper with a new cage to explore. If obstacles to full participation are too great for a child, assign a record-keeping job or lab assistant role, such as keeping track of materials that might be misplaced.

Make use of available professional consultants for specific ways to maximize the potential of a child who has a handicapping condition. Until such helps arrives, put yourself imaginatively into the child's situation to think of ways to work around the child's limitations and through the child's strengths. Whatever accommodations you make to provide satisfying science experiences can help children with special needs grow in ability to cope with the challenges that they face. Information on science activities designed for youngsters who are visually impaired or physically disabled is available from the Center for Multisensory Learning, Lawrence Hall of Science, University of California, Berkely, CA 94720.

Children who face difficulties in mastering developmental tasks deserve every chance to develop confidence and initiative that teachers can provide through science activities. A young teacher, Jane Perkins, did this for her class of learning-disabled children. She was deeply moved by their responsiveness to the science activities. She said, "It was the first time any of these children ever expressed curiosity. I'll never again deny them the chance to wonder why."

LEARNING IN THE CONTEXT OF COOPERATION

Current research on children's thinking and learning regards them as results of both individual development and social interaction. There is also growing recognition in our society of the divisive outcome of emphasizing individual accomplishment and competitiveness in the classroom, while failing to teach children how to work cooperatively. Together, these trends provide impetus to returning to cooperative education. The discovery experiences in this book have always been presented primarily as activities to be shared by small groups, with the introductions, supplements, and follow-ups provided as whole-group experiences. They are readily usable in the cooperative learning context.

Management of discovery activities as cooperative learning tasks will vary according to the nature of the project, the availability of materials in quantity, and the experience of the youngsters in handling independent activities (Johnson, Johnson, Holubec, & Roy, 1988). Certainly, it makes more sense to have only one group of four children working on a particular gravity experience that requires using the one available commercial balance and weights, with other groups working on problems suitable for homemade balance scales. Still other groups could be using that time to make entries in their topic record booklets or to record their earlier findings on a class bar graph. Other children could be working on sections of the class mural of a gravity-free fantasy or creating a skit about being astronauts.

Precisely how cooperative activities are implemented is a matter for the individual teacher to decide. Surely it is wise to make a slow transition to cooperative learning groups from a lockstep approach where each child works independently at the same learning task. Typically, it is realistic to plan a science period where discovery activities are paired with less structured activities that require little supervision. Teachers can then devote maximum attention to facilitating the discovery activities. There is no particular merit in attempting to simultaneously juggle a number of differing discovery activities.

Time must be allowed before the science period ends for cleanup and for processing with the whole class what has occurred in the cooperative groups. To ensure that each child in the class is given a regular opportunity to report what his or her group has accomplished, a number-rotation system can be used. Each child in a small group is assigned a number from 1 to 4. Then, instead of asking for a volunteer or choosing a high-visibility child to report for the group, the teacher indicates which number will identify the day's reporters.

INTEGRATED CURRICULUM THEMES

The science themes in Part 2 of this book were developed in response to the persistent interests, questions, and concerns of a generation of children in the classroom, at play, and at home. To be precise, many of the themes evolved in response to Michael, the gifted child in my first kindergarten class in a university laboratory school 25 years ago. The retentiveness of his mind was impressive. The range of his curiosity was boundless.

Keeping Michael's curiosity satisfied became a major preoccupation for me, searching for resources that weren't to be found, so had to be improvised. Initially, I fretted about devoting too much preparation energy to meet one child's needs. But something quite unexpected was happening in the classroom. Michael was never alone at the science table; the activities always drew a crowd. Soon, during our group conversations, Michael wasn't the only child sharing interesting information about how things work. Michael wasn't the only one applying those learnings in spontaneous play or in solving practical schoolgrounds problems. Others were chiming in to point out connections between story plots and current science projects. Others were bringing from home various artifacts that related to ongoing science learnings.

By midyear, my weekly planning centered on a single new or continuing science theme. We sang science, we nibbled science, we quantified science activities, we danced science, we took science to the easels and workbench, and we trudged around the school's neighborhood to see science happening. We had a good time, the children and I, learning together.

Happily, Michael and his classmates had inspired a science-oriented kindergarten program that continued to evolve over the years. It continued to fascinate children of early childhood ages who represented a wide range of intellectual abilities and family backgrounds. Eventually, the pages of weekly plans and observational notes resulted in the thematic format of this book. Themes were developed around basic science concepts that have functional value in the world of young children because they explain how and why familiar events occur. Some topics were developed to modify or replace the naïve, magical explanations that children had formulated for themselves. Topics were selected according to the following criteria:

- relevance to the child's immediate experience
- potential for being understood through simple, hands-on activities
- potential applicability to practical, child-level problem-solving
- significance in promoting safety and physical or emotional well-being
- availability of low- or no-cost materials for experimentation and observation

Two topics that interest many youngsters, prehistoric animals and astronomy, do not meet these criteria for most early childhood settings. They can be more suitably explored by parents and teachers who can provide for children the resources of a natural history museum, a planetarium or observatory, and the night skies.

Early childhood classes may have opportunities to participate in community environmental projects. Every class can incorporate reusing, recycling, and conserving resources into its daily routines (Earthworks Group, 1990). Such involvement allows children to have a valued role in significant matters, but it does not take the place of foundation science experiences. Basic understandings of how matter behaves in the natural and physical worlds give meaning to the more complex interrelationships of environmental issues. In the chapters that follow, many hands-on experiences are suggested that lead to broader understandings. The activities are sequenced to provide foundation concepts for subsequent activities.

Other worthwhile science topics with child appeal can be developed by creative teachers, as long as the conceptual purpose and scientific accuracy are clear. The hands-on activities must lead children to a general truth or law about how matter and living things behave. Related extensions in other curricular areas should be just as enjoyable as the active science experiences, but they, too, must have a conceptual purpose. They must not be vaguely connected, aimless "things to do" that trivialize the science learning.

A teacher's own fascination with an area of knowledge can be a valid starting point for developing a science topic, because that interest and enthusiasm communicate a positive disposition toward science. The adult's range of knowledge must then be stepped down to the children's level of thinking development and to the context of the child's immediate world. Activities then can be designed around the "up close and personal," tangible aspects of that topic. A teacher's commitment to aerobic exercise and optimal health, for example, can translate into children's beginning experiences with the awareness of air being pulled into their nostrils and filling their lungs; their own pulses and beating hearts, their own moving muscles. The discovery activities could include exploring how breathing rates change and measuring how long it takes to recover normal breath rates after a sprint. A teacher's strong commitment to protecting the environment can translate into beginning, hands-on explorations with reused plastic bottle terrariums and then expand onto the playground and into the broader community. Bess-Gene Holt (1989) offers excellent leads to exploring the immediate environment and toward understanding the patterns and harmony in the undisturbed natural world.

The activities for the science topics in this book are described as one way of supplying accurate guidance for children's explorations. They are sequenced from the simplest concepts to the more complex. The activities illustrating more complex concepts could be made available as options for children who are eager to pursue more challenging ideas. Each teacher is in the best position to make this decision. Children are best served and core concepts are more durably retained when teachers offer only a few topics in depth and with cross-curricular reinforcement each school year than when many topics are touched on lightly.

OBJECTIVES AND ASSESSMENT

The broad objectives underlying the learning experiences in this book focus as much upon the individual's feelings and approach to learning as they do upon the acquisition of information. Observations of the basic harmony in physical and natural world relationships build feelings of security and confidence in children. Reducing fear of the unknown leads to feelings of mastery and strength. Support for curiosity promotes problem-solving ability and offers an avenue for learning how to learn.

In order to quantify certain learning outcomes, behavioral objectives are still used in many schools. However, what a child really learns from an activity might not immediately be evident in the classroom. Instead, that feedback may spill out to a parent at home during a sleepy bedtime conversation. The learnings may be fragmentary until they are consolidated through participation in the integrating activities. Or, perhaps the learning is applied nonverbally weeks after the classroom experience has taken place. The real nature of the teaching/learning process is obscured if we assume that children learn only and precisely what a narrow behavioral objective decrees.

In the following chapters, learning objectives are stated for the teacher's use in making curriculum choices. The objective statements include the science processes used in the activities. *Children are not expected to fully absorb objective concepts in their abstract form.* To avoid detracting from the broad goals of the cognitive/affective approach to science learning, objectives are not stated as behavioral criteria for evaluating the child's performance

Formal evaluation of children's accomplishments in discovery science is required in many schools. It marks our maturity in science education that the difficulty of valid, comprehensive evaluation is widely acknowledged (Hein, 1990). One form of assessment most consistent with discovery science and most useful for identifying misunderstandings that need further clarification is observation of children's verbalizations. Listening to the child's comments and asking questions of the child as she or he works with science materials can reveal a great deal. Such questions include, "What did you find out about. . . ?" and "Can you show me what happens when. . . ?" For longer plant, animal, or weather observations, the question "What did you find out about. . . ?" asked at a natural closure point, provides a gauge for conceptual learning. Children's responses can be recorded most conveniently on a simple checklist. Ideally the check sheet items should center on the grasp of basic concepts, activities completed, and evidence of interest and satisfaction expressed by the child. Such data also serve as self-evaluation for the teacher about the success and value of the learning experiences.

As we increasingly appreciate how the child builds knowledge, partnering with a child in assessing that learning becomes desirable. We may, for example, invite the child to represent her knowledge: "How can you show what you learned about _____?" A journal entry, a drawing, a story, or a dance may result, as well as a verbal explanation. A work folder or a portfolio for that child would include her representation or a description of it. "What else do you wonder about _____?" is a question that further assesses the distance the child has traveled.

The word *assess* originates in the Latin *assidere* meaning "sit by." When we ask children to tell us, in their own way, what they know and what else they would like to know, we do sit by them, facilitating, rather than prescribing, what they learn. Nor should we think that individual assessments are the only way of seeing what children have learned. Just as children work in small groups to develop knowledge, they can present their work jointly. Older children may paint a mural or dramatize a concept. Younger children might engage in creative movement together. Teachers can jot down notes on such activities, as well as photograph or sketch them.

In addition, group conversations about a science topic can both reveal and help develop children's understanding, as Rhoda Kanefsky (1995) has demonstrated. With her first-grade children seated in a circle around her, she introduces a topic and offers each child a chance to comment or ask a question. Writing rapidly, she preserves their words. By the next day she has their remarks typed and posted by the door for all to notice. Meanwhile, interesting topics have spurred children's research, such as talking to their parents or reading books about the topic. This leads to further group conversations, journal writing, and drawing.

The rules of the conversations are simple: no shouting out, no telling anyone that she or he is wrong, and no arguing. Reviewing several conversation transcripts shows how individual children come to correct and refine their thinking, and how they help one another with new ideas and questions. This kind of documentation of children's thinking also helps the teacher decide what to teach next, tightly linking curriculum and assessment (Chittenden, 1990).

Paper-and-pencil tests of factual knowledge for older primary-grade children can be a barrier to adequate responses, since the flow of ideas can be blocked by the need to write, spell, and punctuate correctly the written answer. This problem can be avoided to a certain degree if questions can be asked and answered using simple sketches as well as words.

Assessing science learning is complex. Not only is there much to assess, including concepts, skills and attitudes, but there is also much uncertainty about how science education should be evaluated. Price and Hein (1994) advise:

> Remember, there is no universally accepted, absolute standard for judging the science activities of children. We simply don't have enough information or research evidence to determine what is an "appropriate" or "outstanding" observation for a 6-year-old, or what level of experimental design can be expected from a 10-year-old. (p. 27)

For now, our best procedure may be to use multiple ways of looking at children's work and their involvement in the process.

Teachers make assessments of the whole group as well. How engaged are the children? How well do they keep track of the science materials? How much investigation of the library books occurs? What gets brought in from home or elsewhere outside the classroom? Judgments of these kinds help teachers evaluate their own effectiveness and monitor their own work.

The assessment method most appreciated by parents at conference time, however, does not involve convenient check marks on a rating scale. Rather, it consists of sharing your casual, on-the-spot jottings in a pocket notebook about their child's intense involvement in a fascinating activity or insightful contributions to a class discussion. A collection of these annual notebooks also sums up the teacher's significant contributions to the lives of children. They form a quiet record of having provided children lasting satisfaction in mastering new ideas. Such notes are tangible evidence of having passed along life-enriching awareness of the world we know and appreciate through science knowledge.

GAINING FROM OUR MISTAKES

Many of the materials and activities in this text have been revised as a result of trial-and-error encounters with the logic of young minds. On one occasion, the children in my class were given matching plastic vials, some capped, some uncapped, to use in a buoyancy experience. Then 5-year-old Greg explained why his capped, empty vial floated on the water, while his uncapped vial sank. According to Greg, the cap held up the first vial! To my dismay, he verified his conclusion by removing the plastic cap and floating it on the water. How confusing! According to *my* plan, he should have noticed that the capped vial was filled with air, hence, it was lightweight; the uncapped vial filled with water, hence, it became heavy. It took a while for my supposedly flexible, adult thinking to find a way back to the objective of the experiment. "Greg, you had a good idea about the cap. It does float by itself. Let's see what happens if we put the cap on the container full of water. Now let's put the same kind of cap on the empty-looking container and watch them again." Tuning in to a child's logic makes teaching an exciting learning process for the teacher.

If children have difficulties with an activity, it may be possible that further modifications are needed. It is important not to give up on science activities because of an occasional unexpected outcome. Keep in mind Thomas Edison's observation that a mistake is not a failure if we learn how not to do it next time.

PROFESSIONAL GROWTH

Teaching can be an endlessly challenging profession when we consider ourselves learners along with the children. Taking time to investigate questions we are interested in; talking with other teachers about common concerns in science teaching; refreshing ourselves by attending lectures, visiting museums, and reading can all help to keep us open to new ideas. We can enliven our interest in teaching science by acquainting ourselves with these periodicals for children and for teachers:

- *Connect*   K–8 Hands-On Science and Math Across the Curriculum (bimonthly), Teachers' Laboratory, P.O. Box 6480, Brattleboro, VT 05302
- *Discover*   The lively format of this periodical on science for the nonscientist will keep you posted on new developments. (Subscriptions: 1-800-829-9132)
- *Science & Children*   A journal of the National Science Teachers Association, 1742 Connecticut Ave. NW, Washington, DC 20009
- *Your Big Backyard*   for preschool children; *Ranger Rick*, for primary/elementary grade children; *Ranger Rick's NatureScope*, for teachers (includes activities, extensions, and outdoor projects). These are all publications of the National Wildlife Federation, 1412 16th St. NW, Washington, DC 20036–2266.

The teaching framework will achieve its greatest effectiveness when it stimulates teachers to continue growing and generating their own ideas for presenting science experiences to the children they know best.

REFERENCES

ADLER T. (1991). Cumulative learning aids memory. *American Psychological Association Monitor, 22,* 10.

BLAKE, S. (1993). Are you turning female and minority students away from science? *Science and Children, 30,* 32–35.

BRUNER, J. (1961). *The process of education.* Cambridge, MA: Harvard University Press.

BUTTS, D. P., & HOFMAN, H. (1993). Hands-on, brains-on. *Science and Children, 30,* 15–16.

CAREY, S. (1986). Cognitive science and science education. *American Psychologist, 41,* 1123–1130.

CASE, R. (1986). *Intellectual development: Birth to adulthood.* New York: Academic.

CHITTENDEN, E. (1990). Young children's discussion of science topics. In Hein, G. (Ed.), *The assessment of hands-on elementary science programs* (pp. 220–247). Grand Forks, ND: North Dakota Study Group on Evaluation.

CSIKSZENTMIHALYI, M. (1990). *Flow: The psychology of optimal experience.* New York: Harper & Row.

DWECK, C. S. (1986). Motivational processes affecting learning. *American Psychologist, 41,* 1040–1048.

EARTHWORKS GROUP. (1990). *50 simple things kids can do to save the earth.* Kansas City: Andrews & McMeel.

FARNHAM-DIGGORY, S. (1990). *Schooling.* Cambridge, MA: Harvard University Press.

GARDNER, H. (1991). *The unschooled mind.* New York: Basic Books.

HEIN, G. (ED.). (1990). *The assessment of hands-on elementary science programs.* Grand Forks, ND: North Dakota Study Group on Evaluation.

HOLT, B. G. (1989). *Science with young children.* Washington, DC: National Association for the Education of Young Children.

JOHNSON, D. W., JOHNSON, R., HOLUBEC E., & ROY, P. (1988). *Circles of learning: Cooperation in the classroom.* Alexandria, VA: Association for Supervision and Curriculum Development.

KANEVSKY, R. (1995, March). *Alternative assessment: A developmental perspective toward understanding children's work.* Workshop presented at the National Science Teachers Association National Convention, Philadelphia, PA.

KOHN, A. (1993). *Punished by rewards.* Boston: Houghton Mifflin.

KYLE, W., BONNSTETTER, R., McCLOSKEY, S., & FULTS, B. (1985). What research says: Science through discovery—students love it. *Science and Children, 23,* 39–41.

MORGAN, M. (1984). Reward-induced decrements and increments in intrinsic motivation. *Review of Education Research, 54,* 5–30.

PRICE, S., & HEIN, G. (1994). Scoring active assessments. *Science and Children, 32*(2), 27.

ROWE, M. B. (1974). Relation of wait-time and rewards to the development of language, logic, fate control: Part II, Rewards. *Journal of Research in Science Teaching, 11*(4), 291–308.

PART TWO

CONCEPTS,
EXPERIENCES, AND
INTEGRATING ACTIVITIES

Plant Life

They feed us, clothe us, shelter us, purify the air we breathe, and fill our visual world with beauty: the living things called plants. Children may be captivated by towering giant plants or the tiniest weed blossoms underfoot. When we share their delight, we renew our own appreciation of nature's exquisite order. The following concepts will be explored in this chapter:

- There are many kinds of plants; each has its own form.
- Most plants make seeds for new plants.
- Seeds grow into plants with roots, stems, leaves, and flowers.
- Most plants need water, light, minerals, warmth, and air.
- Some plants grow from roots and stems.
- Some plants do not have seeds or roots.
- Many foods we eat are seeds.

The first suggested experience will be limited by climate to areas where deciduous trees grow. The next group of activities calls for gathering natural materials. This may require the ingenuity of teachers in urban schools. The concluding experiences with seeds and plant growing should be possible anywhere. Suggestions for seedling care and for transplanting are included.

CONCEPT: **There are many kinds of plants; each has its own form.**

1. Do the parts of different plants look different?

LEARNING OBJECTIVE: To observe and describe similarities and differences in the leaves, bark, and flowers of different plant sources. (Do this after taking a walk to collect nature materials.)

FIGURE 4–1

MATERIALS:

Large collecting bag full of
found items such as:

   leaves (2 or 3 of each
     kind)

   tall grasses (include
     seeds or blossoms)

   flowers

   twigs

   bark

   seed pods, nuts

   mosses, lichens

Paper lunch bags

GETTING READY:

Sort the found materials.

Fill a teacher's bag with
one of each kind of item.

Distribute an assortment
of materials into small bags
for the children.

Place a closed bag at each
place at the science table.

SMALL-GROUP ACTIVITY:

1. Take one object from your bag. "Look into your
   bags to see if you can find a leaf that matches this
   one."
2. As children find similar items to compare, encourage them to notice details: "Is it just like mine? Are
   the tips of your leaf rounded like this one? It almost matches. Who found one with rounded tips?"
   (Children may have lots of information about
   plants already. Listen.)
3. Point out that all leaves from the same kind of tree
   have the same general shape. (Size and fall coloration may vary.) For example, "When they have
   finished growing, all sweet gum tree leaves have
   five points. That's one way to know that the tree is
   a sweet gum tree."

Note: Specimens can survive close inspection by enclosing them in a no-cost display cover.
Use a drop of glue to attach a well-dried specimen to the inside of a clear plastic deli-carton
or yogurt container lid. Cover with a matching lid. Join rim edges to form a case and seal
with tape (Figure 4–1). (Remove dating ink by warming the lid over a cup of steaming liquid, then wiping with nail polish remover.)

2. How do some plants rest for winter?

LEARNING OBJECTIVE: To observe and describe seasonal changes in trees and shrubs.

MATERIALS:

Shopping bag

Old, thick catalog

Newspaper

Waxed paper

Electric iron

GETTING READY:

Try to do this a week in advance. Save waxed leaves for another year.

Get a good specimen leaf from each tree you plan to visit with the class.

Press leaves for several days between pages of newspaper inserted into the catalog. Weight with the iron or bricks.

Fold waxed paper over each dried leaf. Press warm iron on paper to coat leaves with wax.

Tape the leaves to a low bulletin board and label.*

SMALL-GROUP ACTIVITY:

1. When deciduous trees start to change color, take a tree trip. Have children circle a tree, holding hands. "What can you see above you? below?" Repeat with an evergreen tree. (Old needles on the ground have been replaced by new ones, but not all at once.) Visit shrubs if trees are not within walking distance.
2. Supply these ideas, as needed: Leaves make food for trees to grow; green color (chlorophyll) in leaves uses the energy from sunlight to help turn water, minerals, and air into food; this work is finished for leaves on some trees when summer ends.
3. Gather leaves from the ground.
4. Let children put like kinds of leaves together when you get back. Ask them to try to match their finds with the mounted specimens. Encourage them to bring leaves from home to try to match them.
5. Save the surplus leaves for art activities or for compost.

Group Discussion: Recall the fun of collecting leaves. Ask what happens to the leaves from deciduous trees and shrubs that rest for winter. Introduce the idea that when chlorophyll goes out of these leaves, other colors that are also in the leaves show instead of only the green chlorophyll. (You might be able to find mottled leaves that fell before all the chlorophyll left.) Talk about how the leaves can still be valuable after they fall. If possible, start a compost bag or heap (see p. 72).

CONCEPT: Most plants make seeds for new plants.

1. Are there seeds in fruit?

LEARNING OBJECTIVE: To become aware that most plants form seeds which may grow into new plants of the same type.

*Other methods of preserving leaves are detailed in *Science and Children,* September, 1994, 32.

MATERIALS:

Any available seed pods:

> flower, tree, shrub, tall grass seed head

As many of these as can be brought in:

> apple, tomato, pomegranate, peach, ear of corn in husk, orange, nuts in shells, melon, squash, green beans, apricot, cucumber

Clean meat trays

Paring knife *(adult use only)*

Smocks

Newspapers

Nutcracker, if needed

GETTING READY:

Wash fruits and vegetables.

Cover table with papers.

Have all children wash their hands.

Put out plants with seed pods first to minimize confusion.

SMALL-GROUP ACTIVITY:

1. Show a seed pod. Open it. (Pods still attached to stalks are best.) Let children tell what they know about it. "Each kind of plant has a job to do: to make seeds for new plants just like it. Each plant forms seeds when its flowers stop blooming. Some seeds are protected by covers we like to eat. Let's try to find seeds inside these fruit and vegetable covers."
2. Take plenty of time to decide if seeds will be inside each fruit. Look at stems and blossom ends. Cut the fruits open and share tastes and sniffs with everyone.
3. Save melon and squash seeds. Later let children wash them and dry them on trays. Save for bird feeding trays. If corn is fresh, pull back husks and hang to dry. Let children shell dried corn for the birds and for a spring sprouting project. Tack one dried husk and ear of corn to a tree for the birds.
4. Encourage older children to figure out a way, such as grouping, to count the large number of seeds in a pumpkin.

Group Discussion: When discussing this experience, consider how many seeds each fruit or vegetable has; that a new plant could grow from each one. Comment on the abundance of nature: a welcome thought for children who hear frequently and worry about endangered species.

Read *The Apricot ABC* by Miska Miles. This is a lovely poem about the cycle from fruit, to tree sprout, to fruit. Try to read it without referring to the alphabet letters hidden in the charming illustrations. The story line may fade in importance if children concentrate on hunting for the letters. This book is no longer in print, but it may still be on library shelves.

2. How are seeds scattered?

LEARNING OBJECTIVE: To examine how different kinds of seeds are formed and scattered.

MATERIALS:

Locally available seeds from garden plants; weeds (teasel, milkweed, burdock); grasses (wheat, oats); and trees (ailanthus, locust, oak, pine,* chestnut)

SMALL-GROUP ACTIVITY:

1. Arrange materials on trays. Encourage children to shake seeds from pods, brush hairy seeds against the mitten, beat grass seed spikes against trays to release grains, and take a close look at burrs and hairy seeds with the magnifying glass to see tiny hooks on the tips.

*Seeds lie beneath separate cone scales. Old cones on trees may no longer contain seeds. Tightly closed new cones will dry and open in a warm oven with the heat turned off. Seeds can then be found.

Tanisha studies the sweet potato roots, leaves, and stems.

Magnifying glasses

Old, fuzzy mittens and socks

Trays

GETTING READY:

Gather ripe seed stalks in advance.

Store in open containers or hang in tied bunches to dry.

Preserve husks and pods intact.

Find weeds in vacant lots or in roadside ditches.

2. Take an envelope of winged seeds and a few heavier nuts to the playground. Let children launch them from a high place. Compare what happens to each kind of seed.

Group Discussion: Talk about how different seeds came out of their coverings in different ways. "Have you ever seen plants growing in places where people couldn't plant them— in sidewalk cracks or rock crevices? How could seeds get there? How many ways can we think of?"

CONCEPT: Seeds grow into plants with roots, stems, leaves, and flowers.

1. What is inside a seed?

LEARNING OBJECTIVE: To examine and notice the parts of seeds.

MATERIALS:

Dried lentils, lima or navy
  beans

Peanuts in shells

Desirable if available: maple
  tree seeds, avocado, fresh
  green beans or peas

Magnifying glass

GETTING READY:

Soak seeds overnight in
  enough water to cover.

Keep a few seeds dry for
  comparison.

LARGE-GROUP ACTIVITY:

1. Comment, "There is a surprise inside a seed. Let's
   find it." Carefully slip off a seed coat. Pull apart the
   two parts (cotyledons) that supply food for starting
   a new plant. Find the "surprise": the tiny new
   plant ready to start growing (embryo).
2. Let the children continue to open seeds and find
   new plants. Offer the magnifying glass for closer
   inspection.
3. If a very ripe avocado is available, slice the fleshy
   part in half. Twist slightly to pull apart. Examine
   the seed halves. Peel the seed coat at the base. A
   ripe fruit seed may already be splitting, revealing a
   root tip. Do *not* split it open.

To Start an Avocado Tree: Slice away about 1/4" (1 cm) from the base of the avocado seed. Insert
three round toothpicks midway through the seed. Suspend the seed in a jar of water. Keep in a
warm place away from direct sunlight. Change the water weekly. After the root appears, move
the jar to a sunny location. When leaves appear on the stem, gently plant in a pot at a depth of
approximately 5" (13 cm). Leave the top quarter of the seed exposed above the soil. Water at
least every other day. Spray or wash leaves frequently. (See Figure 4–3 on p. 61.)

2. How do seeds start to grow?

LEARNING OBJECTIVE: To observe initial seed sprouting.

MATERIALS:

Method 1

Matching disposable plastic
  tumblers

Transparent tape

Cotton balls

Dried legumes: navy or lima
  beans, lentils (fresh stock)*

Water

Plastic prescription vial

Desirable: mung beans (natural food store)

SMALL-GROUP ACTIVITY:

Method 1

1. Make a sprouting dome: Wet 4 or 5 cotton balls,
   press out excess water, and flatten. Line the bottom
   and sides of one tumbler with wet cotton.
2. Let children see and feel the beans. As the children
   help place 4 beans between the cotton and the tumbler side, recall the suprise inside the seeds. (Try to
   use more than one kind of legume to see which
   sprouts first, which gets tallest, and which grows
   for the longest time.)
3. Upend the matching tumbler on the rim of the prepared tumbler. Tape the rims to make a dome enclosure (Figure 4–2).
4. Place it away from direct sunlight where temperature will be even.
5. Put one of each kind of seed in the vial for later comparison. Start a calendar record of starting date, first
   root, first stem, and first leaf appearance. Crayon-
   mark daily growth level on the tumbler.

*Two common causes of germination failure are old seeds and an overheated, dry room. Don't expect goods results with
seeds of unknown vintage, nor with uncovered sprouting containers.

Method 2

Plastic sandwich bags

White paper toweling

Stapler

Masking tape

Method 2

1. Each child makes a sprouting bag. Fold toweling to fit bag and dampen the toweling. Place 5 lentils on damp toweling; staple bag shut. Label with child's name on a piece of tape.
2. Children keep daily sprout growth logs by sketching or writing descriptions, and by measuring root and stem length.

Note: It's a good idea to start two germinating domes. Keep one available for children to pick up for a close look. If sprouts don't survive the inspection, the other dome will be available. It's hard to only look when leaves are showing beneath a rakish seed cover cap.

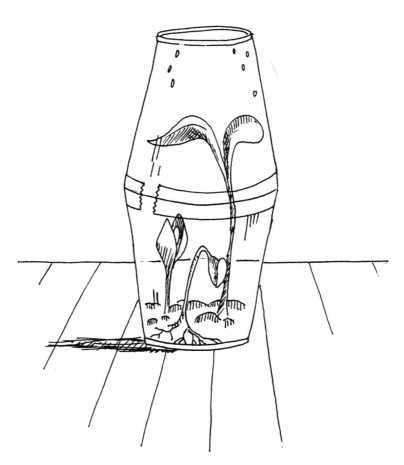

FIGURE 4–2

3. Which direction do roots and stems grow?

LEARNING OBJECTIVE: To observe the tendency of roots to grow downward toward water and of stems to grow upward toward light.

MATERIALS:

Same as for the seed sprouting experience

SMALL-GROUP ACTIVITY:

1. "Notice which direction the seedling roots and stems take. Is it the same for each seedling?"
2. Gently turn one seedling so that the stem points down and the root reaches up. Mark an X on the glass beneath it.
3. Check each day for changes in root and stem growth. Look at the cotton behind the seeds. Roots may poke down into it toward water.

Group Discussion: Ask children if trees grow with their branches and leaves in the soil and their roots in the air; if flowers blossom underground, or if plants send roots into the ground and other parts into the light? Why is this so? Help the children recall that leaves need light and air to perform their food-making job. Roots have the job of getting water and minerals from the soil so that the plant can live and grow. The upended seedling root and stem twisted and turned to grow in the directions where each could get what it needed.

CONCEPT: Most plants need water, light, minerals, warmth, and air.

1. Can we raise plants from seeds?

LEARNING OBJECTIVE: To become aware of and to provide the elements needed to promote seed germination and plant growth.

MATERIALS:

Zinnia or marigold seeds, package-dated for current year

Small package of commercial potting soil (sterilized)

Teaspoons

Eggshells (as large as possible), one for each child, plus spares

Frozen pie pans with plastic dome cover intact*

Water

Sand or fine gravel

Punch-type can opener

Toothpicks

Medicine droppers or spray bottle

SMALL-GROUP ACTIVITY:

Let the children:

1. Fill eggshells almost to the top with soil (cup shell in one hand while filling).
2. Push shell gently into the sand-filled pie pan. Pans can hold about 12 shells.
3. Use dropper, or spray bottle, to dampen soil well. Place one seed on soil, then cover with spoonful of soil, press firmly, water again, and insert name tags in shells. Plant seeds in extra shells to replace possible failures.
4. After all shells are in place, tape plastic domes to pie pans. Place in a spot away from drafts, radiators, and direct sunlight.
5. As soon as tiny shoots appear, remove plastic covering. Move to sunny spot. Keep soil damp with spray or dropper.
6. Start a calendar record of seedling growth stages.

*If frozen pie containers are not available, use two egg cartons, one inside the other. Slide cartons into plastic bags and fasten ends with rubber bands.

GETTING READY:

GETTING READY:

Collect eggshells in advance.

Cover table with newspapers.

Put an inch of sand in each
  pie pan.

Make tiny paper name tags
  and staple to, or lace with, a
  toothpick.

7. Set aside a seedling dampening time every other
   day. Ask, "If you were a tiny new plant, how
   would you want to be cared for?"

Seedling Care: After leaves appear, provide moderate light, such as a northern exposure. Cover only at night to retain moisture. Allow children to water their own plants with the dropper. It's hard to overwater this way, but if it happens, blot up standing water with absorbent materials. Check each shell before covering pans at night in case someone forgets to water a plant.

Transplanting: For several days after the second pair of leaves appears, give seedlings a few hours of direct sunlight, preferably outdoors in a sheltered spot. Pack the shells in small milk cartons stuffed with crumpled paper for children to carry home. Tell parents that the eggshells can be planted directly into the ground if the soil is warm. Crushing the shells with thumb and fingers as the seedlings are placed in the soil helps roots become established.

If many children in the class are apartment dwellers, you may want to grow dwarf plant varieties. Transplant the seedlings before sending them home. Use any container large enough to hold about three cups of soil. Punch a drainage hole in the bottom of cartons or cans, add a layer of small rocks, and fill 1/3 full with soil. Crush the shell, plant firmly, cover with more soil, and water it.

About Soil: To control one possible source of failure, commercially prepared potting soil is recommended for germinating seeds. For other classroom use, let children enjoy mixing their own potting soil: 1/3 ordinary soil, 1/3 sand, and 1/3 peat moss. Talk about how soil is made of crumbled rocks, dead plants, and insect matter.

Group Discussion: Compare the calendar records of the germination dome and the plant-raising experiences. "Were seeds treated alike in both experiences? Which seedlings stopped growing and withered? Which ones kept growing?" Moist seeds can grow only until their built-in food supply is used up. Plants rooted in soil can use minerals and moisture from the soil to help the leaves make their food for growth. Would the seeds grow in a freezer? Find out.

Additional Experiences: Keep some soil on hand all year to be ready for planting opportunities. Plant seeds used in the earliest seed investigation activities. Fresh peas or beans can be planted directly into the soil. Corn must be dry enough to be pried off the cob without breaking open. Plant some pumpkin seeds after Halloween. Try growing plants from grapefruit seeds. Late season, tree-ripened grapefruit seeds seem to respond most promptly. Soak them in a small amount of water until the seeds sink to the bottom of the container (about three days) before planting. Try planting fresh date pits. A general rule for planting is that seeds should be planted at a depth twice the width of the seeds.

2. How do plants take up water?

LEARNING OBJECTIVE: To observe how moisture is taken into the stem of a plant.

MATERIALS:

2 stalks of celery, with leaves

2 jars

Food coloring, blue or red
(enough to make dark color)

Water

SMALL-GROUP ACTIVITY:

1. "How does water get into the leaves of a plant?"

2. "Let's see if we can figure out how water moves up these stalks of celery." Let children stir food coloring into one jar of water.

3. Check within an hour for signs of color in leaf tips. Separate a dyed tube from the stalk so the dye can be seen in the whole length. Slice a cross section from the bottom of the stalk to examine.

4. "What do you think would happen to a stalk of celery if it had no water for a while? Let's find out." Leave the other stalk in the empty jar overnight. Check its condition the next day.

5. "Do you think water will change this stalk? Let's try it." Add water to the jar. Check it the next day. Has it revived? Clarify that celery plants have roots in the ground when they are growing. Roots take up water from the soil and the water travels up through the stalk tubes.

CONCEPT: Some plants grow from roots and stems.

1. Can plants grow from potatoes? carrots?

LEARNING OBJECTIVE: To observe a different means of developing a new plant: by sprouting roots or stems.

MATERIALS:

Part 1

White potato, or homegrown sweet potato. (Shipped ones are treated to prevent sprouting.)

Onion

Clear plastic jar

Water

Round toothpicks

Part 2

Fresh carrot (old one won't do much)

Flower pot saucer

Sand or vermiculite

LARGE-GROUP ACTIVITY:

Part 1

1. Open a discussion by asking children's opinions about whether plants can be grown from parts of the stem or root. Add that potato plants can start this way. "Let's see if we can start one growing in water. We can all watch it in the days ahead to see what happens." (This is a good time to have children draw pictures of this beginning stage.)

2. Stand, or suspend from toothpicks, a potato (sweet or white) in a jar of water. (About 1/3 of the tapered end of a sweet potato needs to be in water.) Add to or change water as needed during the growing period. After the first sprouts appear, ask for children's ideas about what feeds the vine. (Roots and tubers provide stored food for new growth.)

3. Keep a written calendar or a series of observational drawings to record the plant's growth. A successful sweet potato vine will flourish for months before its food supply is depleted. It may be planted in the ground when the old potato starts to cave in. It can still produce a new crop of potatoes.

FIGURE 4–3

GETTING READY:

To avoid disappointing the class, first try at home to sprout a potato from the same source.

Part 2

1. Open a similar discussion another day: "Do you think this carrot top could feed a small new plant?"
2. Fill the saucer with the sand or vermiculite. Dampen it well with water. Cut a 1-1/2" piece from the top end of the carrot. Stand it in the damp sand with the cut end down (Figure 4–3). "Let's keep it damp, and put it in a dark, warm place for a few days to see what happens."
3. As soon as new, green growth appears, move the saucer to a sunny location. Add needed water as long as the ferny plant continues to grow. (This plant will *not* grow into a carrot if placed outdoors in soil. Carrots only grow from seeds.)

Try to plant some daffodil bulbs outdoors in fall. Later, if you should find a sprouting onion in your kitchen, slice it open vertically so the children can see the new plant tucked inside its food supply bulb. Talk about the bulbs outdoors waiting for the spring sunshine to warm the earth and for the rain to start them growing. Try to start a few paper narcissus bulbs or an amaryllis bulb in the room. Follow package directions for growing conditions. The amaryllis bulb is huge and produces spectacular flowers. Children can readily see the flower's pistils and stamen without a magnifying glass. A large seed pod forms as the blossom dies away. It can be opened when dry to provid more seeds for experimenting.

CONCEPT: Some plants do not have seeds or roots.

1. What is a mold?

LEARNING OBJECTIVE: To observe the ways simple plant forms grow and develop.

Group Discussion: Bring a piece of bread, two screw-top jars, and a drop-top bottle (like a soy sauce bottle) of water to a group gathering. Recall with the children that most plants make new plants from seeds or from root and stem parts. Say that a few kinds of plants grow from tiny dust-speck bits called *spores.* Millions of spores blow about in the air, but they are too small to see. When spores land on a warm, moist food source, they grow into plants that we can see.

"Perhaps we can grow a plant like that on a bit of bread." Put half of the bread in each jar. Cover one jar. Let the children sprinkle water on the piece of bread in the second jar. Leave this jar uncovered for an hour. Then cover it and store it in a warm, dark place for a few days. Compare the two pieces of bread. Do they look the same? Leave the moldy bread in the jar to develop a luxuriant fur coat. (Perhaps the black spore clumps will be visible as specks on the mold.)

If you want to develop this concept further, experiment with the growth requirements of yeast. Examine mushroom gills, watch for algae and lichen growing on rocks, sample Roquefort or bleu cheese, watch for brown or green algae growth in an aquarium, bring in mosses and ferns to plant in a terrarium. The green spore-producing plants can make their own food. The colorless spore plants must use dead plants or other sources for food.

Read *Lots of Rot*, by Vicki Cobb, to teach children how molds help reuse the materials in the ecosystem. More experiments with molds and decomposers can be found in *Biology for Every Kid* by Janice Van Cleave.

CONCEPT: Many foods we eat are seeds.

1. Which seeds are good to eat?

LEARNING OBJECTIVE: To experience firsthand that many seeds are used for food.

MATERIALS:

For grinding:

Blender

Cracked wheat

Peanuts

Peppercorns

Pepper mill

For cracking:

Peanuts in shells

Whole sunflower seeds

For toasting:

Hulled sunflower seeds

Hulled pumpkin seeds

Squash seeds

Electric skillet

For cooking:

Rice

Oatmeal

For popping:

Popcorn, salt, margarine

Transparent dome popper

For smelling:

Whole seed spices like nut
  meg, corriander, anise, pop-
  pyseed, or peppercorns

Desirable: pepper mill, mortar
  and pestle, nutmeg grater

SMALL-GROUP ACTIVITY:

Grinding: To show how grain becomes flour, follow blender instructions to grind wheat. Use blender book recipe to make peanut butter.

Toasting: Toast hulled sunflower and pumpkin seeds in an electric skillet at low heat with a few teaspoons of margarine. (**Safety Precaution:** If children are watching, keep well out of their arms' reach.) Stir the seeds until lightly browned. Sprinkle with salt. Delicious and nutritious.

Cooking: Follow package directions. Compare dry and cooked foods, then eat.

Popping: Mention the bit of moisture in dry kernels that changes to steam, exploding the kernels when heated. Plan carefully to ensure safe use of electrical equipment in the classroom.

Sprouting: See page 118.

Read *The Popcorn Book* by Tommie dePaola.

Group Activity: Open a coconut, one of the biggest seeds that grows. Puncture soft spots at the end of it and drain the juice into a measuring cup; then let children sample spoonfuls of juice. Crack the shell with a hammer and scrape out small pieces of coconut for children to taste. If a blender is available, grate some of the coconut according to manufacturer's instructions to make this confection:

SEED CANDY

Use the blender to grate:	1 cup of coconut
	1 cup hulled sunflower seeds
Mix with:	2 tablespoons peanut butter
	2 tablespoons honey, confectioners sugar, or maple syrup (both of the latter come from plants)

Form into a log, slice thinly, and enjoy.

Go on exploring other foods from plants as long as your enthusiasm lasts. Try, or talk about, root, stem, blossom, or leaf parts of plants that we eat.

Safety Precaution: Caution children that some growing things are poisonous. Eating or tasting such things could make them extremely sick. They should never pick or eat wild mushrooms. Find out what the children already know about this subject. Do not bring the following poisonous berries into your classroom: castor beans, berries from pokeweed, nightshade, bittersweet, yew, holly, privet, or mistletoe. Do not allow experimental eating of buckeyes, daffodil bulbs, iris or lily tubers, or poinsettia leaves. Further information about poisonous plants can be found in *A Guide to the Medical Plants of the United States* by A. and C. Krockmal.

PRIMARY INTEGRATING ACTIVITIES

Math Experiences

Keeping Records. Take a shopping bag on nature walks to hold things found along the way. Use these things in the classroom for sorting, classifying, and counting experiences. Record the results on a bulletin board or an experience chart (*We Found*   12 leaves, 10 acorns, 8 bits of bark, 1 piece of moss). For younger children, use numerals and samples of the material; for beginning readers, use numerals and words.

Accumulate Collections. Chestnuts, acorns, pinecones, sweet gum "balls," buttonwood "buttons," and sticks can be collected. Use them as materials to sort, match, count, and weigh. Use them as markers in bingo games. Group them on meat trays to make sets to match with numeral cutouts or numerals marked on cardboard squares. Arrange sticks in order according to their length.

Recording Growth. Use string to record plant growth. Hold the string beside the plant stem. Cut string at the level of the stem height. Tape the strings each day to a posterboard chart, and label each string with the date to keep a visual record of the plant's growth.

Classifying Seeds. Give children a jar of varied dried beans and legumes to sort into ice cube trays. Have children describe differences and similarities of the seeds.

Estimating Change. Show children the measureful of popcorn called for in corn popper directions. Ask children how many measuresful of popped corn they think this amount of seeds will make. Write down their estimates. Use the same measure to count the popped results, with everyone counting aloud as a group. Make another batch, again asking for estimates. Some children will be ready to revise their estimate based on the previous counting. Some will stick with their original idea.

Estimating Large Numbers. For older children, give small groups pint jars of beans. Ask for estimates of the number of beans. "How could we find out?" Actual counting is an acceptable way. Children may want to count out a row of beans, then count the number of rows, or devise other shortcut ways to avoid laborious piece-by-piece counting. For large numbers, encourage the use of calculators to simplify the repeated adding or multiplying involved.

Music (Resources in Appendix A)

Try these children's songs about plants.

1. Sing "The Tree in the Woods," the old English cumulative song about a tree, its parts, and its occupants.

2. Sing from *The 2nd Raffi Songbook:* "In My Garden" (the digging, planting and picking could be pantomimed as well), "Popcorn," "All I Really Need" (the rain and the sun to give life to the seeds), and have fun singing "Apples and Bananas."

3. Mary Miche sings "Dirt Made My Lunch" on her cassette, *"Nature Nuts."*

4. Dan Crow sings "The Zucchini Song" on his cassette, *"A Friend, A Laugh, A Walk in the Woods."*

5. Sing the traditional "Oats and Beans and Barley," *Readers' Digest Children's Songbook.*

Plants as Music Makers. Dried gourd maracas are well-known plant instruments. Drums or resonant rhythm instruments can be made from hollow logs and coconut shells. Don't forget willow whistles, dandelion or grass stem whistles, wooden flutes, and recorders. Try to show a set of wooden tone blocks.

Stories and Resources for Children

ARDLEY, NEIL. *The Science Book of Things That Grow.* San Diego: Harcourt Brace Jovanovich, 1991. This book is logically organized to present plant growth basics through challenging questions, and provides carefully detailed experiments to find the answers. Handsome layouts and color photographs invite involvement.*

BRANDT, KEITH. *Discovering Trees.* Mahwah, NJ: Troll, 1982. Development from seed to tree, plus seasonal aspects of the growth cycle, are narrated at preprimary child's interest level. Paperback.

BRANLEY, FRANKLYN. *The Sun: Our Nearest Star.* New York: Harper & Row, 1988. Includes experiments demonstrating that energy from the sun helps plants grow. Paperback.

BUNTING, EVE. *Flower Garden.* San Diego: Harcourt Brace, 1994. Outstanding, warm illustrations enrich a child's story of how she prepares a flowerbox surprise for her mother.

BURNIE, DAVID. *Plant.* New York: Alfred A. Knopf, 1989. Beautiful color-photograph layouts and clear information in this book for slightly older audiences will fascinate primary-grade children as well. Also see in this Eyewitness Series by the same author, *Tree,* 1989.

BURNIE, DAVID. *Flowers.* New York: Dorling Kindersley, 1992. Fresh perspectives in the groupings of flowers from many habitats. Includes suggestions for using flowers to make art and craft projects such as pressing, drying, and making potpourri.*

CARLSTROM, NANCY. *Wild Sunflower Child Anna.* New York: Simon & Schuster, 1991. Sun-drenched watercolors by Jerry Pinkney enrich poetic descriptions of Anna reveling in the wonders she finds in the bright morning meadow. Paperback.

COBB, VICKI. *Lots of Rot.* New York: J. B. Lippincott, 1981. The mildew plant is shown as a nuisance *and* a necessity. Amusingly illustrated.

*Starred references, written at the young child's level of understanding, can also help teachers with minimal backgrounds in science to expand their knowledge base.

COOMBES, ALLEN. *Trees.* Eyewitness Handbook. New York: Dorling Kindersley, 1992. This comprehensive guidebook gives identifying characteristics of more than 500 species of trees.

DAMON, LAURA. *Wonders of Plants and Flowers.* Mahwah, NJ: Troll, 1990. A broad look at plant life, from the microscopic to the gigantic. Upper primary-grade children. Paperback.

dePAOLA, TOMIE. *The Popcorn Book.* New York: Holiday House, 1978. Two children acquire interesting lore about this favorite food as they prepare a huge batch for themselves. Paperback.

EDWARDS, MICHELLE. *Eve and Smithy.* New York: Lothrop, Lee, & Shepard, 1994. Two vintage neighbors share gardening and art in this pleasant, nicely illustrated story.

FOREY, PAM. *Wild Flowers of North America: Science Nature Guides.* San Diego, CA: Thunder Bay Press, 1994. Color illustrations and basic information identifying 180 wildflowers by habitat. Includes directions for drying flowers and flower art projects for children.

GIBBONS, GAIL. *From Seed to Plant.* New York: Holiday House, 1991. Engaging illustrations and focused text accurately inform youngsters about the plant growth cycle from seed to seed.

HELLER, RUTH. *The Reason for a Flower.* New York: Grosset & Dunlap, 1983. Strikingly beautiful, fascinating illustrations by the author, and a simple text make this a book that children will return to and remember.

HOWARD, ELLEN. *The Big Seed.* New York: Simon & Schuster, 1993. A satisfying story for preschoolers in which the smallest girl in the class grows the tallest plant. Lillian Hoban's illustrations show Bess's stepdad stir-frying vegetables for dinner.

JEUNESSE, G., DELAFOSSE, C., & METTLER, R. *Flowers.* New York: Scholastic, 1992. Excellent acetate overlay pages add depth by uncovering the inner working parts of the blossom to intrigue preschool/primary children. Also in this First Discovery series is *Trees,* by the same authors.

JORDAN, HELENE. *How a Seed Grows.* New York: Harper Collins, 1992. An easy-to-duplicate bean-growing experience is described in simple text for young preschoolers.

LOBEL, ANITA. *Alison's Zinnea.* New York: Greenwillow, 1990. Striking flower paintings are the focus of this clever, alphabet-sequenced story line. Its alliterations are fun to read and hear.

LOBEL, ARNOLD. *Frog and Toad Together.* New York: Harper & Row, 1972. Toad has some funny ideas about coaxing seeds to grow. Paperback.

LOBEL, ARNOLD. *The Rose in My Garden.* New York: Morrow, 1993. A cumulative tale in lovely prose that takes place among garden flowers. Charming illustrations by Anita Lobel. Paperback.

MARKMANN, ERIKA. *Grow It! An Indoor/Outdoor Gardening Guide for Kids.* New York: Random House, 1991. Complete growing guides with special attention to year-round container gardening, as well as directions for growing a tent, making perfume, and more. Delightfully illustrated, this book won gardening awards in Germany.*

MILES, MISKA. *The Apricot ABC.* Boston: Little, Brown, 1969. This is a beautifully illustrated story of insects' responses to a fallen apricot that eventually becomes a fruit-bearing tree. It is no longer in print, but libraries may still have it in circulation.

MITCHELL, ALAN. *Trees: Science Nature Guides.* San Diego, CA: Thunder Bay Press, 1994. Identifying characteristics of common trees of North America, plus child activities.

MULLER, GERDA. *The Garden in the City.* New York: Dutton, 1993. Charmingly detailed illustrations by the author and a good story line help children absorb gardening information. A nice assortment of activities are tucked into the pages of this pleasant book.

PHILLIPS, SARAH (ED.). *Plants: What's Inside? Series.* New York: Dorling Kindersley, 1992. A simple text, color photographs, and cutaway drawings will help young children understand the function of flowers.

ROMANOVA, NATALIA. *Once There Was a Tree.* New York: Dial, 1989. This lovely ecosystem tale from Russia is enriched by soft, detailed, Durer-like illustrations. New York Times Notable Books of the Year award. Paperback.

RYDER, JOANNE. *Hello Tree!* New York: Dutton, 1991. Use this charming evocation of the bond between a special tree and a child as a stimulus to creative thinking or writing activities.

SILVERSTEIN, SHEL. *The Giving Tree.* New York: Harper & Row, 1964. A boy grows up with a tree that gives him shade, apples, shelter, and finally, a resting stump.

TAYLOR, BARBARA. *Green Thumbs Up!* New York: Random House, 1991. Directions are given for growing plants from seeds, bulbs, cuttings, leaves, tubers, and tops of certain fuits and vegetables. Learn how to improvise the tools to create a bottle garden. Paperback.

TRESSALT, ALVIN. *The Gift of the Tree.* New York: Lothrop, Lee & Shepard, 1992. Soft watercolors enrich the story of a century of shelter and eventual promotion of new life. A fine story of the life cycle of a single oak tree.

WILKES, ANGELA. *My First Garden Book.* New York: Alfred A. Knopf, 1992. Excellent color photographs illustrate directions for planting seeds, growing a tiny desert garden, creating a hanging basket garden, raising plants from fruit seeds, and more.

WYLER, ROSE. *Science Fun with Peanuts and Popcorn.* New York: Simon & Schuster, 1989. Seeds to grow and seeds to eat! This book combines science with the child appeal of favorite foods. Paperback.

WYLER, ROSE. *Grass and Grasshoppers.* New York: Julian Messner/Simon & Schuster, 1990. The text and illustrations move gently from good feelings about lying in grassy fields to wondering, observing, experimenting, and learning about grass and the small creatures it shelters.

Poems (Resources in Appendix A)

The collection *Poems to Grow On,* compiled by Jean McKee Thompson, has four poems about seeds appropriate to read during the seed investigations:

> "The Little Plant," by Kate Louise Brown
> "The Seed" and "Carrot Seeds," by Aileen Fisher
> "Seeds," by Walter de la Mare

X. J. Kennedy's poem, "Art Class," in *I Thought I'd Take My Rat to School* by Dorothy Kennedy (Ed.), describes a child's speculations about drawing a tree.

Fingerplays

This traditional fingerplay fits well with the concept that living things reproduce in their own special form.

The Apple Tree

Way up high in the apple tree	(Stretch arms up.)
Two red apples, I did see.	(Make circles with hands.)
I shook that tree as *hard as I could.*	(Shake "trunk.")
Mmmmm, those apples tasted good!	(Pat tummy.)

—AUTHOR UNKNOWN

Substitute "orange carrots, green pears, two bananas," and so on for "two red apples, I did see." The children will enjoy catching and correcting your mistake. Ask them why it must be apples growing on the apple tree. "Really? Don't carrots grow on apple trees? Then, where do they grow? Do they grow in the ground from apple seeds?"

My Garden

This is my garden	(Hold one hand palm upward.)
I'll rake it with care.	("Rake" with curled fingers of other hand.)
Here are the seeds.	
I'll plant in there.	(Pantomime planting, seed by seed.)
The sun will shine.	(With arms make circle above head.)
The rain will fall.	(Fingers flutter down.)
The seeds will sprout.	(Spread fingers of one hand. Push up other fingers between them.)
And grow up tall.	(Bring hands and forearms together. Move up, spreading palms outward as arms move up.)

—AUTHOR UNKNOWN

Art Activities

Collage. Dried grasses, leaves, pressed flowers, flat seeds, and small twigs make lovely collage materials. Tape may be needed when younger children include twigs and long grasses in their work. Vary the background colors to bring out the hues of the dried materials.

Rubbings. Tape a single fresh leaf or a pattern of small leaves to the table in front of each child. Cover with a sheet of heavy paper. Let children rub a crayon over the paper to bring out the relief design of the leaf veins.

Translucents. Fresh leaves, flower petals, and grasses can be sealed between two sheets of waxed paper to make translucent window hangings. Do not use an electric iron in the classroom to seal the paper. For safety, and for maximum child participation, use a newspaper-covered electric food warming tray as a heat source. Give children a pizza roller or a child-size rolling pin to apply light strokes of pressure to the waxed paper.

Cut-Paper Mural. Flower inventions: Offer strips of green paper for stems and leaves and assorted sizes of circles in blossom colors. Indicate ways to change the texture and shape of the rounds such as notching; slashing for fringes; scoring with a ruler edge and pinching into cup shapes; cutting into spirals; and curling edges around a pencil. Put out scissors and white glue or paste. Assemble the children's fantasy flowers into a beautiful mural for your classroom door.

Leaf Mobile. Let children cut out freehand leaf shapes (or whatever satisfies them as appearing leaflike). In spring, use green paper; in fall, use orange, brown, and red. Use 5-inch squares of paper. Suggest making the job easier by folding the paper in half. Let them punch a hole in each leaf and thread it with a small bit of yarn. Help them tie or tape their leaves to an interestingly shaped branch. Hang the branch from a ceiling beam, or staple it to a bulletin board with some of the twigs extending into the room.

For more nature art ideas, consult *Good Earth Art: Environmental Art for Kids,* by Mary Ann Kohl and Cindy Gainer.

Dramatic Play

Farming. Playgrounds that offer a bit of shady ground for digging are natural settings for spontaneous farm play. Provide children with small, sturdy rakes, hoes, shovels, buckets, and a wheelbarrow. They will find rocks or cones to plant, and leaves, grasses, or pine needles to harvest without further suggestion. Children will need to know the boundaries of the permissible digging area. Listen to their planting ideas.

Sand Table Indoor Gardens. A collection of twigs, pinecones of several sizes, dried grasses and pods, or perhaps blossoms can be arranged by the children to create small landscapes in the sand table. (Dampened sand will hold better than dry sand.) Rubber toy animals and people can be added.

Blossom Fun. If you can find an abundance of grassflowers for children to gather, use the small blossoms to make beautiful (though short-lived) decorations. Show children how to make a split midway in a dandelion stem in which to insert another dandelion stem, and so on. The resulting rope can be looped into necklaces or crowns, or be allowed to get as long as possible. Leaves and sturdy blossoms can be threaded on soft, covered wires from a telephone cable to make bracelets. Large leaves strung together can become headdresses. Two hollyhock blossoms can be stacked and joined through the center with a toothpick to make dolls with hollyhock-bud heads and daisy hats.

Children can place small sprigs of evergreen in a water-filled ring mold to make a tabletop Christmas wreath. Use small pinecones for turkey bodies; tiny hemlock cones, painted red, for heads; bird feathers for tails.

Creative Movement

Curl up on the floor with the children, pretending to be seeds that have been planted in spring (or bulbs planted in fall if this was a class project). Move with the children to enact the growing story as you softly tell it. "Here we are waiting under the ground. The sunshine makes the soil warm; rain falls, and we begin to expand. The tiny plant inside grows bigger and pokes out of the seed cover. We send a root down to get water. Now our stem starts to push its way up . . . up . . . up to find the sunlight." Slowly describe the growth of the plant above the ground: leafing; budding; blooming; swaying in the breeze; feeling the sun and rain; losing petals; forming seeds; then slowly withering and scattering seeds for next year's plants.

Creative Thinking

What If? What if you were as tiny as your thumb? Which plants would you want to live near? to sleep in? to use for food? to hide under when it rains? to climb for fun? Would you enjoy curling up inside a tulip blossom for a nap? For inspiration, read "The Little Land," by Robert Louis Stevenson, in *A Child's Garden of Verses* (Appendix A).

Play a simple guessing game in which children take turns describing mounted pictures of plants for others to guess. "I'm thinking of a plant that we eat for breakfast, makes something for blue jays to eat on the feeding tray," and so on. Seed catalogs are good sources for pictures. (Pictures printed directly on cardboard food boxes do not need further mounting.) Include a piece of plain paper, a mounted ice cream stick, and a cotton ball or fabric.

Ask children to listen with eyes closed, imagining a special tree to enjoy, as you read aloud Joanne Ryder's *Hello Tree!* Afterward let children draw the trees they imagined.

Food Experiences

Make applesauce when apples are in season. Try drying raisins if grapes are abundant in your area. Children can learn to scrub and peel raw fruit or vegetables for snacks and prepare vegetables for soup. Any time flour or sugar is used in classroom cooking projects, mention its plant source.

Discuss the parts of plants that are being served for lunch. "Are we eating the leaf, the root, or the stem of the celery plant (the potato, the carrot)?" Talk about the wheat seeds and the stems of sugar cane plants or the roots of sugar beet plants that went into the snack-time graham crackers.

Field Trips

We usually think of pleasant woods and meadows as ideal sites for plant life field trips, but the closest grocery store also has important plant learnings for children. Go there to see how much we depend on plants for food. Visit the produce section; the shelves of dry staple foods (flour, pasta, sugar, legumes, cereals, and all the packaged mixes); the canned and frozen fruits and vegetables; and the baked goods. A natural foods shop (not a health food store that features vitamin preparations) is a fine place to see what whole grains look like, and to watch a small flour mill in operation. Flower shops and greenhouses can be fascinating, but make sure children are welcomed by the owners.

Indoor/Outdoor Tree Walks. How many different kinds of trees grow within walking distance of your school? How can children tell that they are different? If leaves are too high to be examined closely before they fall, bark characteristics can provide identification clues. Let children make bark rubbings with old crayons on thick, flexible paper. The backs of vinyl-coated wallpaper samples are good for this. Compare the differences in bark textures for different species.

Now move indoors to complete your tree walk. Ask the children to tour the classroom looking for all the wood they can find in use there. Old buildings may have ceiling moldings and window and door frames of wood, so suggest looking up to find wood. Make a list of the found items. Pencils and paper should be on the list, too.

Having a Quiet Look Outdoors. When the weather is good, explore a nearby weedy patch. Minimize the scatter tendency of unconfined children by defining small observation spaces with 6-foot loops of yarn. Before going outdoors decide on groups of three or four children to share a looking loop. Children can stretch out on the ground, radiating from the loop like the spokes of the wheel. Ask children to report what they see in their space. Later make a summary chart of the observations, including soil, rocks, and small creatures, as well as plants. Tape samples to the chart.

Grassflower Gathering. Check grassy areas near your school in early spring for the presence of tiny flowers. Try to obtain permission for your children to pick tiny bouquets to take home. (Small groups make it a happier occasion.) Show the children where they may hunt, then let them take their time to find spring beauties, chickweed, violets, or whatever the lawnmower spares. Put the flowers in capsule vials of water. Label each with the child's name. Use them for table decorations, then wrap each child's flowers in a twist of waxed paper when it is time to go home.

SECONDARY INTEGRATION

Maintaining Concepts

A row of potted plants on the classroom windowsill guarantees year-round attention to plant growth needs. Plant tending can occupy an honored position on the children's daily job chart. Younger children may need some help carrying the pitcher and deciding how much water to use. Older children can handle the task independently if pots are labeled with suitable watering advice. Comment on changes taking place, such as new buds, fading leaves, and unwanted insect tenants.

Classes in session during the regional growing season might be able to keep plant life concepts in focus by planting and maintaining a garden. Other classes could try portable gardens: plants started at school in the spring, moved to the home of a child or teacher for summer care, and returned to school for a fall harvest—with luck. Start pumpkin and sunflower seeds in good soil, using 2-gallon plastic buckets with drainage holes cut in the bottom. Thin out all but one vigorous seedling. Give it full sunshine and plenty of water. The plants won't attain full growth, but the pumpkin can produce a vine with leaves, tendrils, blossoms, and possibly a fledgling pumpkin. The sunflower will develop a seed head and may grow taller than a 5- or 6-year-old. Slip a section of old hosiery or netting over the fading blossom to keep the birds from feasting on the seeds too soon.

If long-term growing projects cannot be worked out, perhaps you can duplicate the efforts of one very fine teacher of 3-year-olds. Each fall she scouts the countryside to find a whole, dried cornstalk complete with ears of corn. It creates a fall harvest mood in her room, and provides an awesome lesson for the children who seem dwarfed by the giant that grew from a single kernel of corn. Later, shelling the corn becomes an absorbing task for the children. They feed some kernels to the birds, and save some to germinate in spring. Other plant life experiences sometimes come about inadvertently. Capitalize on them. For example, an aging jack-o-lantern might develop a moldy spot. Instead of quietly discarding it, keep it in the room (as long as odor permits) to observe the changes. Then add it to a compost pile so that it can contribute soil enrichment for next year's plants. Read *Lots of Rot* by Vicki Cobb.

Connecting Concepts

Soil Composition Relationships. The natural cycle of renewing limited materials in the ecosystem is well-illustrated in the creation of fertile soil. Pulverize shale to form powdery clay (see Chapter 10). Try growing an extra seedling in it. Compare its growth with seedlings growing in true soil that also contains bits of decayed plant and animal matter. This idea could lead to starting a compost pile or bag in the fall. Let the children rake available leaves and grass clippings into a large plastic trash bag. When it is approximately half full, add a soup can full of fertilizer or powdered lime, a can of water, and a few shovelfuls of dirt. Close the bag and leave it outdoors where it won't be forgotten during the winter. It will need to be turned, shaken, and opened several times for air. Although the completeness of the change that will occur by spring is not predictable, bacterial growth within the bag environment will have interacted with the materials to promote decay into compost (humus) to enrich the soil.

If molds appear on some of the decaying materials, examine them with magnifiers. Talk about molds as tiny, leafless plants that use other materials as food. Vicki Cobb's book, *Lots of Rot,* describes the value of molds as decomposers.

Discuss how completely rotted natural materials improve the soil and make it possible for strong, new plants to grow and produce food. Talk about renewing the soil this way as one of the wonderful cycles of nature: from living plants to decomposition . . . from enriched soil to healthy living plants again . . . and again . . . and again. It is comforting to understand the concept that once-living things continue to have a function after they die (see Furman, 1990, in Resources).

Plants That Help Crumble Rocks. Look in shady areas for greenish-gray patches of lichen growing on rocks. Lichen are fungi and algae that live together as a single unit. Together they make acids that slowly dissolve the rock surface.

Other plants sometimes grow in rock crevices, and their roots may break off pieces of rock. Perhaps there is a section of concrete sidewalk near your school that has been cracked or pushed up by strong tree roots. Watch for them as you take walks with children. Stop to look at them. Recall with the children that slowly crumbling rocks become part of the soil that plants grow in. (See Resource, Chapter 10.)

Air/Water Cycle. Use the term *evaporation* when plants are being watered. Some of the water will be taken up by the roots of the plants; the rest will evaporate. (See Resource, Chapter 8.)

If you make a terrarium, talk about how rarely it will need to be watered. Bring out the idea that the water taken up by the air in the closed terrarium will change into large drops on the cool glass sides. If you make the terrarium before the children have tried the evaporation and condensation experiences, postpone the discussion.

To make a small terrarium from a 2-liter plastic soda bottle, begin by pouring hot water into the bottle to soften the glue adhering the base and label to the bottle. Use a table knife to separate the loosened base from the bottle. Cut off the neck portion of the bottle so it will fit, inverted, into the base. (Do these steps away from children.) Put a layer of gravel and some charcoal bits in the bottom of the base.

Add a layer of potting soil. Arrange small plants in the soil. Dampen the soil; do not saturate. Fit the inverted dome firmly into the lid to seal the terrarium.

Family Involvement

Children may take care of keeping their parents well-informed as to the progress of their seedling experiences. Families can be asked to save eggshells for the project. Their help will be needed in providing a good growing location and in overseeing the care of the seedling that is sent home. Printed plant care tips could be fastened to the container. Include a few lines about other aspects of the projects that can help parents become informed listeners to their children.

RESOURCES

BERGER, THOMAS. *The Harvest Craft Book.* Edinburgh: Floris Books, 1992. Includes directions for making the cornhusk dolls and wheatstraw figures that older children could make as a craft for related studies of pioneering life. Available through "Hearthsong: A Catalog for Families," P. O. Box B, Sebastopol, CA 95473.

FURMAN, ERNA. Plant a Potato—Learn About Life (and Death). *Young Children, 46* (Nov. 1990), 15–20.

Growing Ideas. This great, free journal of garden-based learning activities is published three times a year by the National Gardening Association, 180 Flynn Ave., Burlington, VT 05401. It is sponsored by the National Science Foundation. Write for a complimentary issue and subscription form.

HAMPTON, CAROL, & KRAMER, DAVID. *Classroom Creature Culture; Algae to Anoles.* Washington, DC: National Science Teacher's Association, 1994. This anthology of articles from *Science and Children* on the care of plants and animals brought in to school from the wild is an important resource for teachers.

KOHL, MARYANN, & GAINER, CINDY. *Good Earth Art.* Bellingham, WA: Bright Ring Press, 1991. Contains many good art extensions of plant life.

LERNER, CAROL. *Moonseed and Mistletoe: A Book of Poisonous Wild Plants.* New York: Morrow, 1988.

PETERSON, ROGER TORY. *Wildflowers.* (Northeastern & North Central America). Boston: Houghton Mifflin, 1986. Color-coded page corners speed identification of specimens in this pocket-size field guide. Paperback.

PETRASH, CAROL. *Earthways.* This gentle book includes directions for braiding wheatstalk wreaths and making simple cornhusk dolls, time-honored crafts from pioneer days.

RUSSELL, HELEN ROSS. *Ten-Minute Field Trips: Using the School Grounds for Environmental Studies.* (2nd ed.), Washington, DC: National Science Teacher's Association, 1991. Every city-bound teacher should know this book. Nature's ability to triumph over asphalt and concrete permeates the text.

VAN CLEAVE, JANICE. *Biology for Every Kid.* New York: Wiley, 1990. Includes simple experiments with bean seedlings and molds.

ZIM, HERBERT S., & MARTIN, ALEXANDER. *Trees: A Guide to Familiar American Trees.* A Golden Nature Guide. New York: Golden Press, 1987. Inexpensive, comprehensive pocket guidebook.

Animal Life

Animals of all sizes and conditions fascinate many children who are eager to watch, touch, and care for creatures. Other children have limited tolerance for anything that creeps, crawls, or nips. The experiences suggested for this chapter can both expand the knowledge of the creature lovers and soften the feelings of anxious children into moderate respect for the useful and beautiful small animals around us. These concepts will be explored:

- There are many kinds of animals.
- Animals move in different ways.
- Each animal needs its own kind of food.
- Many animals make shelters in which to rear their young.

The feasibility of this chapter's experiences will depend upon having specimens for observation. Flexible planning is necessary. It would be sheer luck for a spider and a teacher to meet precisely on the day set aside for spider study.

Buying, housing, and maintaining classroom pets can be expensive. However, earthworms, spiders, and insects exist everywhere and can be easily obtained. The experiences that follow use insects, worms, fish, wild birds, and pictures to illustrate concepts. Suggestions for acquiring and understanding insects, as well as ideas pertaining to borrowed pets, are included.

CONCEPT: There are many kinds of animals.

Introduce this topic with a question for children to think about: "What is an animal?" Responses about specific animals will flow easily. Then suggest that there are so many kinds of animals in the world that it would take days just to say their names. "Here is a shorter way to say what an animal is: *An animal is any living thing that is not a plant.*" How many animals can the children think of now? Their list can include people, spiders, earthworms, and insects. There are more than 800,000 species of insects alone! All insects have some features in common.

1. What is an insect?

LEARNING OBJECTIVE: To discover ways in which insects are similar to each other.

MATERIALS:

Temporary cages (see p. 77)

Live insects

Preserved insect in plastic
  container (optional*)

Magnifying glasses

Paper

Crayons

GETTING READY:

Follow capture techniques
  that follow.

Keep live insect cages out of
  direct sunlight.

SMALL-GROUP ACTIVITY:

1. Ground rules for observation should be set:
   (a) Insects or other small animals stay in the cages
   and (b) cages must be handled gently.
2. Suggest looking for things that identify members
   of the insect family: three body parts (head, thorax,
   abdomen); six legs; two feelers (antennae). Spiders
   are not insects (8 legs); caterpillars *are* (only 6 legs
   are true, jointed legs).
3. Encourage children to try drawing pictures of the
   insect they are watching.
4. Mention: (a) These are adult insects. First they
   were eggs, then larvae (wingless, wormlike) before
   changing to adult form. (b) Insects have no bones.
   They have stiff coverings protecting their soft bodies.
5. Release insects outdoors at the end of the day.

Further observation activities are detailed in *The Pillbug Project* by Robin Burnett, and *Pet Bugs: A Kid's Guide to Catching and Keeping Touchable Insects* by Sally Kneidel. (See Resources, p. 100.)

Capture Techniques

Locating Creatures. Start looking for specimens near your own doorstep. On warm nights check screen doors for insects that are attracted by light. Hunt for web-building spiders on window frames and shrubs. Turn over rocks on the ground to find crickets, beetles, and such. Look in flower borders for ladybugs, bumblebees, grasshoppers, ants, and wandering spiders that chase their prey instead of catching them on webs. Examine weed clumps like Queen Anne's lace and milkweed for caterpillars.

Catching Creatures. Cold-blooded animals do not move rapidly in cool parts of the day. A bumblebee is groggy and easy to catch early on a crisp fall morning. Scoop it up with an open jar, then quickly clap on the lid. Slip in a dewy sprig of the plant that the bee rested on and re-cover the jar with a piece of nylon hosiery or netting, held with a rubber band. Bees may be easier to locate during warmer parts of the day *but are quicker to defend themselves.* Try using bamboo toast tongs to catch an alert bee. Use tongs to pick up very spiky-looking caterpillars. Some cause skin rashes if touched.

Wandering ground spiders and grasshoppers may be caught by clapping an open jar over them. Both species seem to hop straight up inside the jar, making it easy to slip the lid under it.

---

*Youngsters with a strong fear of insects may shrink from examining live specimens. They may be able to approach a mounted insect, safely sealed in a plastic container.

Nets are preferred for the safe capture of butterflies and moths. Sweep the net over the insect; flip the bag to fold it over the catch. Remove the butterfly by gently holding two wings folded back together. Slip it into a sandwich bag. Keep the insect out of the sun until it can be transferred to an observation terrarium or box.

Only two small spiders have a dangerous bite. Avoid:

- *The black widow:* Shiny black body with a bright red hourglass mark beneath the abdomen (back section). Young may have three red dots on top of the abdomen.
- *The brown recluse:* Rare. Lives indoors in attics, closets, or other dark corners. Yellow to brownish color. Fiddle-shaped mark on top of front section (cephalothorax) is dark brown. The base of the fiddle is between the feelers; the neck of the fiddle runs toward the spider's abdomen.

More specific advice on capturing and caring for insects is given in Sally Kneidel's fine book, *Pet Bugs: A Kid's Guide to Catching and Keeping Touchable Insects.*

Temporary Housing

There are several ways to make inexpensive cages for small creatures; however, releasing a specimen after a day in the classroom teaches responsible stewardship. Comment that the insect is needed outside for cleanup work or pest control, or to help flowers make seeds (see pages 79 and 99). This also skirts the problem of providing live food for some small animals.

A simple temporary cage for small insects can be made by covering a clean plastic tumbler with a piece of nylon hosiery, stretched and held in place with a strip of tape. For larger insects, cut large windows in the sides of a plastic deli container, as shown in Figure 5–1. Put in a bit of wet cotton for moisture. Pull the foot

FIGURE 5–1

half of an old nylon stocking over the container, gathering the top tightly with a rubber band. Snap the carton lid over the gathering.

Some reusable, clear containers for fragile berries have ventilation holes and snap-on lids. If available, they make perfect cages without any further preparation. Keep cages on hand to use whenever children bring small creatures to share. (It can be difficult to sell some children on the cage idea. Joey, for one, was in favor of keeping his caterpillar on a leash of "gotch tape.") More specific ideas on housing and care for visiting insects are offered in *Classroom Creature Culture: Algae to Anoles*, edited by Hampton and Kramer.

One fascinating exception to the same-day-release rule is a mother spider. She does not feed during the period of egg-sac construction, nor during the weeks before the spiderlings emerge from the sac. The spiderlings also have a stored food supply for use after their birth.

Wolf spiders fasten egg sacs to their undersides. The babies migrate to the mother's back after emerging. There, they cling as passengers to knob-tipped hairs. A wolf spider and her young can survive for a week of observation in a plastic playing-card box if moisture is provided.

A potential disappointment with spider watching is that the mother spider spins an egg sac whether the eggs have been fertilized by a male spider or not. If the eggs are infertile, the mother does not tear the sac open.

Mealworms are easy insects to study. They may be obtained from bait or pet stores very inexpensively. Usually a classroom colony can be supplied for less than $2. They do not bite or fly around alarmingly, being content to spend their lives burrowing in a glass or plastic container of oatmeal or bran about 3" (76 mm) deep. Cover the bran with a slightly damp paper towel, and provide a piece of apple or potato for refreshment. Lift off the towel to observe what is going on.

In the "worm" or larvae stage, mealworms are dry, 6-legged, small, brown, segmented creatures. This stage is interesting to watch for a few weeks, until they seem to "die." Actually, though, they are entering a chrysalis stage, from which they emerge as light-colored beetles about an inch long. The beetles darken as they age, mate, and die in about a week. Their tiny eggs hatch into very small larvae at first, and are almost invisible. The larvae grow quickly into the interesting stage of "mealworms." The life cycle happens reliably, providing surprises for the children. "Where did *these* bugs come from?" For more information, see *Pet Bugs: A Kid's Guide To Catching and Keeping Touchable Insects,* by Sally Kneidel.

Insect Pests

Some insects do more harm than good to humans and plants. Among them are flies, mosquitoes, cockroaches, black widow and brown recluse spiders, miller moths, and clothes moths. Many insects that sting to defend themselves will not bother people if they are not disturbed. Among them are wasps, hornets, bees, and spiders (Table 5–1).

TABLE 5–1 Interesting, helpful insects

Insect	Function	Interesting Features
Ants	Scavengers who clean their surroundings. Newly discovered function: pollinators of flowers.	Social insects who live in colonies with separate jobs to perform: some nurse young, some gather food.
Bees	Highly valued pollinators of plants. Producers of honey and beeswax.	Social insects who live in colonies with specialized jobs to perform.
Butterflies	Pollinators of plants. (Explain to children that they help plants make seeds.)	Slender bodies. Antennae have ball tips. Fly by day. Fold wings straight up when resting. Usually form a chrysalis.
Moths	Some are pollinators. Silk moths spin strong, lustrous fibers that are made into fabric.	Fat, furry bodies. Feathery antennae. Wings spread flat when resting. Usually spin cocoons. Fly at night.
Beetles	Some kinds are scavengers who tidy up. Ladybug beetles are valued by gardeners for pest control. Some beetles destroy crops.	Many handsome varieties: striped, spotted, iridescent.
Crickets	Serve as food for other animals. They are destructive to some crops.	Only male crickets chirp. They raise and rub their hard wing covers to make vibrations.
Fireflies (soft-bodied beetles)	Appreciated for adding charm to summer nights.	Fireflies signal to one another with flashes of light.
Grasshoppers, Locusts	Destroy some crops but serve as food for other animals.	Wings barely visible when flying. Use hind pair of legs to jump. Hearing area is in the abdomen.
Praying Mantis	So valuable for pest control that egg cases are sold to gardeners.	Almost 4″ (10 cm) long, so body parts easily seen. Frightening to see them tearing other insects apart.
Wasps and Hornets	Some wasps pollinate fruit trees. Some eat destructive larvae.	Hornets and some wasps chew dead wood into paper to make nests for their young. Some wasps make nests from mud. Beautiful engineering.

Word recognition comes easily when important facts label the gerbil cage.

2. What is a mammal? (Classifying mammals and nonmammals)

LEARNING OBJECTIVE: To become aware of mammalian characteristics.

MATERIALS:

Part 1

Caged small mammal:* gerbil, mouse, guinea pig

Part 2

Plastic models or mounted small pictures of mammals (include a human), large and small, domestic and wild, and nonmammalian animals (birds, fish, amphibians, reptiles, insects, spiders)

GETTING READY:

Collect small colored pictures of animals. Mount on index cards or attach them to small yogurt carton lids.

Set up YES/NO trays. Provide fur and a picture of a nursing

SMALL-GROUP ACTIVITY:

1. Establish rules for observing, not disturbing, animals. Guide children to notice mammal characteristics:
   a. "How is the animal covered?"
      (All mammals have some hair.)
   b. "Why do its sides move in and out?"
      (All mammals need to breathe air.)
   c. "If we could stroke it, could we feel bones under the fur and skin?"
      (All mammals have backbones.)
   d. "Do you suppose it hatched from an egg, or did it grow in its mother's body before it was born?" (Mammals develop in their mother's body.)
   e. "How do you think this animal was fed as a baby?"
      (Female mammals make milk to nurse their young.)
   f. Ask children if people have hair, bones, lungs to breathe; if they developed inside their mothers; if mothers can nurse their babies? (Humans are mammals.)

*A local pet store or nature center may be a lending resource.

animal as YES tray cues. Provide shells, eggshells, reptile skins, or pictures as NO tray cues.

2. Invite pairs of children to play the mammal classifying game with the pictures or model animals.

CONCEPT: Animals move in different ways.

1. How does an earthworm move?

LEARNING OBJECTIVE: To observe and describe the ways in which different animals move.

Introduction: "Do you need six legs to walk? If your body were bent over and close to the ground, would more legs help keep you balanced? What helps some insects and birds move through the air? Can you think of a very small animal that has no legs—one that only has muscles and tiny bristles to help it move? Perhaps we'll find some worms to watch when we go outdoors."

MATERIALS:

Shovels and a digging place

Produce trays

Light-colored sand

Bucket, stick

Soil

Peat moss

Water

Newspapers

Plastic shoe box

Nylon net, rubber band

Dry oatmeal, coffee grounds, or cornmeal

GETTING READY:

Before taking children out side, check soil—where you have permission to dig—for the presence of earthworms.

If children can't do the digging, bring worms and a shovelful of soil to school.

Cover with a plastic bag. See *How to Hold an Earthworm* in the following section.

SMALL-GROUP ACTIVITY:

1. Let children dig for earthworms, if possible. Give them equal amounts of peat moss and soil in the bucket to mix with a stick. Add water to dampen well.
2. Spread a thin layer of sand on the bottom of the shoe box. Cover with 3″ (8 cm) of mixed soil.
3. Sprinkle meal or coffee grounds on top; scratch in with the stick. Put all but two earthworms on the soil. Cover top with net fastened with large rubber band.
4. Put two earthworms on trays to watch them move. (Bristles are retractable and hard to see. A wide, light band at midsection is the egg case. The tail end is tapered and the head end is rounded. They have no eyes.)
5. See the following section for earthworm farm care. Periodically check sides and bottom of the box for signs of change. Dark soil streaks in the light sand show how earthworms mix soils.

Care of Earthworms

Earthworm Farm Care. Wrap black paper around the sides of the shoe box to encourage tunneling where it can easily be seen. It will take a while for the earthworms to get used to the new soil before they start tunneling. Remove the paper for short periods of time to check for signs of activity.

Earthworms will not survive long in dry soil. Keep the farm away from a heat source. Sprinkle the soil frequently to keep it damp. The box lid can be placed loosely over the net to retain moisture, but some air must enter the box.

Work in a spoonful of dry food each week. Try adding bits of grass and leaves. Tiny balls surrounding tunnel holes on the surface of the soil are the castings of soil digested by the earthworms. This is one way that worms enrich the soil. They also help water and air reach plant roots as they make their tunnels. For more detailed information, see *Wonderful Worms* by Linda Glaser.

How to Hold an Earthworm. Youngsters credit adults with unlimited ability and courage. Measuring up to this idealism can be hard for some of us when it means holding a child's cherished earthworm. Bolster your courage for this eventuality by understanding your tactile senses: Compare the sensation of holding a

A fish, a boy, and a hand lens come together for quiet observation of fins in motion.

slippery object in the palm of your hand with that of holding the same object between your thumb and forefinger. The palm of the hand has fewer nerve endings
to convey tactile sensations. Therefore, a placid worm that is loosely cupped in the
palm of your hand will scarcely be felt.

2. How does a fish move?

LEARNING OBJECTIVE: To observe and describe the ways in which different animals move.

MATERIALS:

1 widemouthed gallon jar

Aquarium gravel or lake sand

Rocks

Water plant—purchased or
from lake or stream

1 small goldfish

Black paper

Newspaper

Water

GETTING READY:

Involve children in as many
preparation steps as you can.

Let a gallon of water stand
overnight in open containers
(so chlorine can escape).

Wash local water plants carefully. Wash lake sand in a deep
pan by letting a slow stream of
hot water fill pan. (Plants and
gravel from a good pet store
need not be washed.)

Put an inch of sand or gravel
in jar. Put plant roots in sand
and anchor with a rock.

Cover sand with folded
newspaper while pouring in
water.

Remove paper.

Put fish in water.

SMALL-GROUP ACTIVITY:

1. Look at the body of the fish. Discuss how it is different from the earthworm's body.
2. Recall the looping, sliding movements of worms.
   Compare with varied movements of the fish as it
   darts up, down, forward, backward, or rests.
3. Find 7 fins: 2 pairs approximately where our arms
   and legs grow, topside, underside, and tail fins.
   (Arm and leg position fins work fast to move fish
   forward and back, top and bottom fins give balance, tail fins steer fish as it swings from side to
   side. Some fins look like fancy decorations; although they have delicate bones, they work hard.)
4. Tap the tank lightly. Does the fish go faster? Does it
   move differently to speed up? Watch closely.
5. To keep the fish as a class pet, put fish feeding on a
   rotating routine chart. Each day a different child
   can feed it a *pinch* of tropical fish food. No weekend feeding is needed. Keep the aquarium away
   from direct sunlight. Tape dark colored paper
   around the side closest to the window to cut down
   algae growth. Some light is needed, however.
   Siphon or dip out one third of the water and replace with aerated water as needed.

CONCEPT: Each animal needs its own kind of food.

Introduction: "Did you have a nice bowl of acorns and a plateful of grass with ladybugs for
breakfast today? Why not? What did you have?" Help children recognize that each kind of

animal needs its own kind of food in order to live. Discuss making a feeder to help winter birds get the kind of food they need. Plan to maintain the feeder until spring. Some birds may come to depend on that food supply and starve without it.

1. Can we feed winter birds?

LEARNING OBJECTIVE: To value helping wild animals find food.

MATERIALS:

Gallon milk carton or clean
  plastic bleach jug

12″ (30 cm) stick

Scissors

Twine

Commercial wild bird seed
  plus seeds saved from plant
  experiences

Dried pinecones

String

Peanut butter

Plastic pint berry baskets

Twine

Suet

SMALL-GROUP ACTIVITY:

A

Cut out two sides of carton or jug, leaving a two-inch border near the bottom to hold seeds. Poke the stick through the feeder in border edges to form two perches. Pierce holes through the carton top for hanging twine, or wind twine around jug top and handle. Tie to a tree branch low enough for easy refilling.

B

Wind string around top scales of cones; make a hanging loop. Let children use spoon handles to stuff peanut butter between scales.

C

Let children lace two baskets together with twine to form a closed container. Add chunks of suet before last side is lashed together. Fasten to tree.

Note: Try to position feeders out of reach of squirrels or cats. Sprinkle seeds on the ground beneath feeders for a few days to attract birds. Check *The Bird Feeder Book* by Donald and Lillian Stokes for further information.

CONCEPT: **Many animals make shelters in which to rear their young.**

1. Do ants care for their young in nests?

LEARNING OBJECTIVE: To observe the social nature of ants.

Introduction "When you were a tiny baby, did your mother put you outdoors to find your own food? Why not? Many kinds of animals make shelters where they can take care of their babies. Perhaps we can see how ants do this." Discuss making an ant farm. If this cannot be done, try to observe an anthill outdoors. Gently poke into the hill with a stick. The children may be able to see nurse ants carrying eggs, larvae, and cocoons as they leave their disturbed nest.

MATERIALS:

Widemouthed quart jar

Slim olive jar to fit in the quart jar

Construction paper

Rubber bands

Piece of nylon hosiery

Bit of sponge

Flashlight

Jam or syrup

Newspapers, trowel, double grocery bag

Ants, nest soil

GETTING READY:

Do this away from children the first time. *You may be very busy for a while.* Locate and dig into nest of medium or large ants. Quickly scoop ants, larvae, eggs, cocoons, and soil into bag. Fasten top well. (Do not mix ants or soil from two nests.)

Keep bag in cool place until needed.

SMALL-GROUP ACTIVITY:

1. Make this ant farm outdoors. Let children center closed olive jar inside the quart jar; fold and tuck newspaper into top of quart jar as a funnel.

2. Gently transfer ants and soil from bag to jar. Quickly stretch hosiery over jar top; slip on a rubber band. Let children wet the sponge and place a dab of jam on it. Quickly slip it into the jar and cover again with hosiery.

3. Let children wrap black paper around the jar, fastening it with rubber bands or tape.

4. Let children keep the sponge damp by dropping water through the hosiery daily; replenish jam weekly.

5. In a few days let children remove paper for a few minutes to see what the ants are doing. A flashlight may be helpful. Lift jar to check activity at the bottom of the jar. Rewrap jar after viewing. Try to locate food gatherers, nurses for young, nest cleaners, soldiers, and hump-backed queen.

2. Can we help nest-building birds?

LEARNING OBJECTIVE: To value helping wild birds provide shelter for rearing their young.

Introduction: "Do you think that a mother bird lays eggs on tree branches and leaves them to hatch by themselves? Of course not! She works hard, sometimes with the father bird, to build a nest where she can keep the eggs warm and keep the hatched babies safe and fed. When the babies are big enough to take care of themselves, usually the whole family leaves the nest and never uses it again. Let's take a close look at a nest." (Hunt for your nest when deciduous trees have lost their leaves. Look in dense shrubs, heavy vines, tree crotches, and under eaves. Ask your county agricultural extension agent for suggestions on finding nests and determining if they are really abandoned, since some species return to old nests each year. Check for exhibits in a library or museum, if you cannot find a nest.)

MATERIALS:

Abandoned bird nest

Clear plastic shoe box

Tape

Plastic berry baskets

6" (15 cm) pieces of yarn

SMALL-GROUP ACTIVITY:

1. Let children examine the nest structure. "Is it lined with special material? Why?" Keep nest in the plastic shoe box, taped shut for easy handling and storage.

2. If another nest is available, let children gently pull it apart to see how hard the birds work, piece by piece, to build a nest. Try to rebuild it.*

*__Health precaution:__ Children should avoid touching their mouths and eyes while handling the nests, and should wash their hands thoroughly afterward.

Spanish moss, if available

Dried tall grass

3. Let children prepare nest materials for birds that come to school feeders. Pull apart clumps of Spanish moss, place loosely in berry baskets, and work some strands through the holes. Weave strands of yarn and long grass through bottom and side holes of other baskets.
4. Hang baskets near feeders.

Read the poem "The Bird Nest" from *Earth Lines* by Pat Moon.

Animals in the Classroom

The Borrowed Pet. Much can be learned by children who help provide the daily life requirements of a classroom pet. Many of those learnings can also be sampled during a short visit by a borrowed pet.

Before the pet arrives, discuss with the children safe ways to watch and care for it. For small pets accustomed to pens, make an observation box from a large carton. Cut windows in the sides and cover them with plastic wrap, if needed. Some pets are better off being held by their owners during the visit. Try to let children see the animals eating and drinking water. Help children find answers to questions about how the animal moves, gets its food, and protects itself. Does the pet have bones inside, or a hard outside covering? Does it have hair or fur, smooth skin or feather-covered skin? Does it nurse its young? Does it build a shelter for its young?

A monarch butterfly rested in our terrarium, inches away from awed observers.

Recent animal visitors to our school included a pet boa constrictor, who laced himself through the rungs of a chair, and a small pony who was unexpectedly led into the building by a mischievous owner.

Animal Rearing. Rearing butterflies or moths from caterpillars, chrysalid, or cocoon stages can be enthralling or disappointing. Try following the procedures described in *How to Raise Butterflies* by Jaediker Norsgaard. Strong interest and luck are required for success.

The same mixture of dedication and luck also contributes to a good outcome with an egg-incubating project. A commercial incubator is probably a better choice than improvised equipment for classroom use. Follow the instructions included with the incubator. Be sure the fertile eggs have not been allowed to cool after being laid. Plan to find a home for the chick or duckling after it has hatched.

Are Classroom Pets Necessary? Good teachers allow for individual differences in children's responses to animals. They do not assume that all children adore animals. Also, they do not insist that a fearful child make physical contact with animals.

Teachers should extend the same consideration to their own feelings about animals. While they should avoid expressing negative attitudes about animals to children, they need not feel obligated to undertake year-long care if they do not enjoy the experience. Neither should they feel inadequate if they prefer to skip the responsibility of animal care. The teacher's primary affective focus and primary responsibility are caring about and caring for children. Feeling similar warmth and commitment toward animals is an asset, but not a requirement.

PRIMARY INTEGRATING ACTIVITIES

Math Experiences

Many commercial math materials incorporate animals:

1. Animal Match Ups are puzzle cards with sets of animals designed to link with numerals.
2. Magnet and flannel board animal counting sets.
3. Sequence puzzles include the development from egg to butterfly; from egg to frog; and from nest building to robin egg hatching. The pictures are to be arranged in frames according to time sequence.
4. Animal Rummy encourages children to make sets of like animals in a game context. A reduced number of the cards can be used for playing Memory (Concentration).
5. *Over In the Meadow,* read or sung, involves animal counting in a literary context. (See Appendix A)
6. Frog Math contains set-making and probability games, spinning off from Arnold Lobel's *Frog and Toad* books. Obtainable form Great Explorations in Math and Sciences (GEMS), Lawrence Hall of Science, Berkeley, CA. (See Appendix A)

Use collections of small seashells as materials for counting, classifying, set making, or numeral game markers.

Keep a flannel board tally of small animals brought into your classroom or observed on walks. Mount pictures of the animals, cut from magazines, on sandpaper or flock-backed adhesive paper backing. Bring out the envelope of pictures for children to count when new observations are reported.

If your class tries a chick-incubating project, make a chain of 21 large paper clips to represent the days of the incubation period. Remove a clip each day and count the days left in the waiting period.

Music (Resources in Appendix A)

Sing along with "You Can't Make a Turtle Come Out" on Mary Miche's cassette, *Earthy Tunes.* It reminds children that, to observe animals, ". . . . you'll have to patiently wait." Also on that cassette is "Snakes and Spiders."

Many traditional and contemporary children's songs complement ongoing animal life study:

"Shoo Fly, Don't Bother Me"
"The Eency Weency Spider"
"Jig Along Home," by Woodie Guthrie (Worms and grasshoppers are among the dancers.)
"Three Little Ducklings"

Listen to *Birds, Beasts, Bugs and Little Fishes,* sung by Pete Seeger. The songs were written by Ruth Seeger in *Animal Folk Songs for Children.* Also listen to "I Like the Animals in the Zoo," sung by Ella Jenkins on *Seasons for Singing.* The song provides a good pattern for improvising about the animals that visit your room:

I like the grasshopper, in the jar.
I like the grasshopper, it jumps far.

Animal horns were among the earliest forms of musical instruments. Conch shells are still used as horns in some parts of the world. One of the percussion instruments on our classroom music shelf is an unoccupied box turtle shell. The children enjoy tapping rhythms on it.

Stories and Resources About Animal Characteristics

BURTON, JANE. *Animals Eating.* Brookfield, CT: Newington Press, 1991. One of a series of books that looks at animal characteristics along a single dimension of functioning. Others are *Animals Fighting* and *Animals Talking.*

COLE, JOANNA. *Large as Life: Daytime Animals,* and its companion, *Large as Life: Nighttime Animals.* New York: Alfred A. Knopf, 1985. This pair of books features life-size paintings of wild animals. The beautifully rendered textures of fur, feathers, and spines are outstanding. The text describes animal characteristics of interest to young children. Nature notes at the end provide supplementary information.

EMORY, JERRY. *Nightprowlers.* San Diego: Harcourt Brace, 1994. This book for older children makes a good general reference about nocturnal creatures both for the classroom and for parents, who are the available night nature guides. Paperback.

EPSTEIN, SAM & BERYL. *Bugs For Dinner? The Eating Habits of Neighborhood Creatures.* New York: Simon & Schuster, 1989. Menus and more about insects and small animals familiar to city children.

LILLY, KENNETH. *Animal Builders.* New York: Random House, 1984. Also in the series: *Animal Climbers, Animal Jumpers, Animal Runners,* and *Animal Swimmers.* Simple concepts about animal characteristics for the youngest preschoolers, printed on durable board for easy handling.

WHITCOMBE, BOBBIE. *Animals in Danger.* New York: Checkerboard Press, 1990. Provides clearly written capsules of information in large type for young readers. Intriguing illustrations are generously used. Also in this series of inexpensive, hardcover books are *Animals of the Night, Dangerous Animals,* and *Strange Animals.*

ZOLOTOW, CHARLOTTE. *Peter and the Pigeons.* New York: Greenwillow, 1993. Peter's first visit to the zoo introduces him to new animals, but his favorite animals are the familiar pigeons.

Stories and Resources About Insects, Spiders, and Earthworms

BERNHARD, EMERY. *Dragonfly.* New York: Holiday House, 1993. Careful illustrations with a light touch reflect the clear, complete text describing this ancient, fascinating, and benign insect.

BERNHARD, EMERY. *Ladybug.* New York: Holiday House, 1992. Good information on the life cycle, usefulness, and folklore about this beetle is easy to absorb in this nice blend of art and science.

BOOTH, JERRY. *Big Bugs.* San Diego: Harcourt Brace, 1994. *Big* refers to the size of this book, not the bugs. Written for older children to "take the ug out of bug," the facts are accurate and activities lively. Paperback.*

CARLE, ERIC. *The Very Busy Spider.* New York: Philomel Books, 1985. A spider's workday is revealed in this engaging story for preschoolers.

CARLE, ERIC. *The Very Quiet Cricket.* New York: Philomel Books, 1990. A cricket's search for its voice helps young children respect some insects with whom we share the earth.

CRAIG, JANET. *Amazing World of Spiders.* Mahwah, NJ: Troll, 1990. Fascinating information about the diversity of spiders and the intricacy of their work. Paperback.*

DORROS, ARTHUR. *Ant Cities.* New York: Thomas Y. Crowell, 1987. The organized life beneath the anthills is carefully presented for young children.

FELTWELL, JOHN. *Butterflies and Moths.* Eyewitness Explorers. New York: Dorling Kindersley, 1993. Great color photographs enlarge children's understanding of these creatures.

GIBBONS, GAIL. *Spiders.* New York: Holiday House, 1993. Simple language and bold, clear illustrations distinguish spiders from insects and demonstrate the ingenuity used by various spiders as they trap their food supplies.

GRAHAM, MARGARET BLOY. *Be Nice to Spiders.* New York: Harper & Row, 1967. A fine story about a garden spider that improves life for the animals in a zoo. It promotes respect for spiders.

HAWCOCK, DAVID, & MONTGOMERY, L. *Ant.* New York: Random House, 1994. Small, simple, and to the point, this is a rare pop-up book that will survive the classroom. The ingenius three-dimensional model is designed to be suspended overhead by a cord attached to its covers. Neat! Also in this Bouncing Bugs series: *Bee and Spider.* These are "books that turn into bugs."

HELLER, RUTH. *How to Hide a Butterfly.* New York: Platt & Munk, 1992. The clever, accurate rhyming text is minimal. Combined with the author's fascinating, precise illustrations, this book about protective camouflage in nature makes an unforgettable impact. Paperback.

*Starred references, written at the young child's level of understanding, can help teachers with minimal backgrounds in science to expand their knowledge base.

LASKY, KATHRYN. *Monarchs.* San Diego: Harcourt Brace, 1993. This photo-story of the miraculous life cycle of the monarch butterfly includes a description of the annual children's butterfly parade celebrating the migratory return of the monarchs to California.

MOUND, LAURENCE. *Amazing Insects.* New York: Alfred A. Knopf, 1993. Insects looked at with the fresh perspectives, striking color enlargements, amusing cartoons, and intriguing facts usually found in the Eyewitness Juniors series. Paperback.

MOUND, LAURENCE. *Insect.* New York: Alfred A. Knopf, 1990. Beautifully organized information and sparkling color photographs. Eyewitness Series.

NORSGAARD, E. JAEDIKER. *How To Raise Butterflies.* New York: Dodd, Mead, 1988. Exact color photographs enrich this book that takes the guesswork out of rearing butterflies in the classroom.*

PARKER, STEVE. *Insects.* Eyewitness Explorers. New York: Dorling Kindersley, 1992. Provides interesting perspectives for understanding insects. It describes making and maintaining an insect aquarium. Paperback.

PARSONS, ALEXANDRA. *Amazing Spiders.* New York: Alfred A. Knopf, 1990. Children will become better observers because of the simple text and exceptional illustrations in this Eyewitness Junior book. Paperback.

PHILPOT, LORNA & GRAHAM. *Amazing Anthony Ant.* New York: Random House, 1994. Follow a jaunty ant through pages of clever mazes while singing endless verses of "The Ants Came Marching in a Line." A flip tab feature provides clues and lures for language acquisition. Information about ant predators is imaginatively incorporated into the illustrations.

PLACCO, PATRICIA. *The Bee Tree.* New York: Philomel (Putnam & Grosset), 1993. Grandfather uses a bee tree search, and the prize of honey, to encourage Mary Ellen to search for the sweetness of knowledge in books.

PORTE, BARBARA. *Leave That Cricket Be, Alan Lee.* New York: Greenwillow, 1993. A captured cricket doesn't make a sound inside the jar Alan Lee provides for its home. Set free, it settles down and sings again for this Asian-American boy.

ROYSTON, ANGELA. *Insects.* New York: Dorling Kindersley, 1992. Clever illustrations seem to peel away the external view of common insects to reveal the internal workings. Intriguing information presented in straightforward style. What's Inside Series. Paperback.

VAN ALLSBURG, CHRIS. *Two Bad Ants.* Boston: Houghton Mifflin, 1988. This imaginative story of adventuresome ants can spark children's creative thinking and perspective-taking.

WHITCOMBE, BOBBIE. *Insects.* New York: Checkerboard Press, 1990. Fascinating facts and carefully detailed illustrations invite young readers.

WYLER, ROSE. *Grass and Grasshoppers.* New York: Simon & Schuster, 1990. This book will teach you how to ease an earthworm out of its hole, and how to trap and cool down a grasshopper for closer observation.

Stories and Resources About Fish, Mollusks, Amphibians, Reptiles

ABBOT, R. TUCKER. *Seashells.* Science Nature Guides. San Diego: Thunder Bay Press, 1994. 175 color photographs identify shells from four shore areas in the U.S. and Canada.

ALIKI. *My Visit to the Aquarium.* New York: Harper Collins, 1993. Fascinated children are caught up in the sea life exhibits surrounding them. Facts about marine life are easy to absorb in this good story.

BELL, SIMON (ED.) *What's Inside Shells?* New York: Dorling Kindersley, 1994. Vivid color photographs of shelled creatures are paired with cross-sectional drawings of the snails, oysters, tortoises, crabs, and other mollusks living within. Eyewitness Explorers series. Paperback.

BERGER, MELVIN. *Watch Out for Turtles!* New York: Harper Collins, 1992. An easy narrative style inspires children to respect this creature. NSTA Outstanding Science Book Award.

CARLE, ERIC. *A House for Hermit Crab.* Natick, MA: Picture Book Studio, 1988. A wonderful blend of accurate portrayal of ocean floor life with a metaphor about the pangs of growth and change in children's lives.

COLE, JOANNA. *Hungry, Hungry Sharks.* New York: Random House, 1986. Describes many kinds of sharks, from hand-size dwarfs to giants. Reassures that sharks do not go hunting for people. Paperback.

COLE, JOANNA. *The Magic Schoolbus on the Ocean Floor.* New York: Scholastic, 1992. Mrs. Frizzle's "learning while laughing approach," along with a class equipped with air tanks, makes complex information about undersea life both intriguing and palatable.

DALY, KATHLEEN. *The Golden Book of Sharks and Whales.* New York: Golden Books, 1989. Sharks, the biggest fish, and whales, the biggest mammals, seem a bit less fearsome with this interesting information and graceful art.

FINE, JOHN. *Creatures of the Sea.* New York: Atheneum, 1989. Fascinating color portraits and descriptions of unusual sea dwellers.

GIBBONS, GAIL. *Frogs.* New York: Holiday House, 1993. Simple graphics and text follow the life cycle of the frog from egg to tadpole to adult.

JAMES, SIMON. *Dear Mr. Blueberry.* New York: Simon & Schuster, 1991. In this story for younger children, imagination, ingenuity, and curiosity lead a girl to simple information about whales.

KOHN, BERNICE. *The Beachcomber's Book.* New York: Viking Press, 1987. Describes many of nature's gifts from the sea, and how to use them in children's crafts and cooking.

MAZER, ANNE. *The Salamander Room.* New York: Alfred A. Knopf, 1991. A conversation between a child and his mother is captured in lovely prose and illustrations to imaginatively inform us about the life requirements of a salamander.

McCLOSKEY, ROBERT. *One Morning in Maine.* New York: Viking Press, 1952. Clam digging and shore animal life are part of this classic story.

MILTON, JOYCE. *Whales and Other Sea Creatures.* New York: Random House, 1993. Dramatic things go on in the dark ocean floor among sharks, squid, and eels; but in this informative portrayal, the whale moves along benignly. Paperback.

PARKER, NANCY. *Frogs, Toads, Lizards and Salamanders.* New York: Greenwillow, 1990. Whimsical jingles and illustrations accompany the scientific descriptions and drawings of amphibians. A good retention device.

PARSONS, ALEXANDRA. *Amazing Snakes.* New York: Alfred A. Knopf, 1990. Vivid illustrations and a text designed to intrigue young readers. Paperback.

RYDER, JOANNE. *Lizard in the Sun.* New York: Morrow, 1990. Ryder's lovely prose invites us to experience a fantasy day as a proud, small lizard, changing color, basking in the sun, finding food. Beautiful illustrations by Michael Rothman make the adventure vivid. Paperback.

RYDER, JOANNE. *One Small Fish.* New York: Morrow, 1993. Imagination turns a dull moment in the science room into an undersea adventure for a girl. Wonderful sea creatures swim out of the pages and crawl through backpacks as readers explore marine life with the girl.

RYDER, JOANNE. *Snail in the Woods.* New York: Harper & Row, 1979. A satisfying story that reveals the nature of the common snail.

SABIN, LOUIS. *Reptiles and Amphibians.* Mahwah, NJ: Troll, 1985. A good source of answers for older preschoolers. Paperback.

SMITH, TREVOR. *Amazing Lizards.* New York: Alfred A. Knopf, 1990. Vivid illustrations and a text designed for young readers. Eyewitness Juniors.

WHITCOMBE, BOBBIE. *Whales and Sharks.* New York: Checkerboard Press, 1990. Generous illustrations and large type to engage young readers.

WINNER, CHERIE. *Salamanders.* Minneapolis: Carolrhoda Books, 1993. The detailed information offered on these fascinating amphibians can be summarized for young children. Excellent color photographs. NSTA Outstanding Science Tradebook. Paperback.

Stories and Resources About Birds

BURNIE, DAVID. *Bird.* New York: Alfred A. Knopf, 1988. Beautifully organized information and sparkling color photographs. London Natural History Museum Eyewitness Series.

CURRAN, EILEEN. *Bird's Nests.* Mahwah, NJ: Troll, 1985. The very simple text is intended for preschoolers, but the lovely, precise illustrations will catch the attention of older children also.

GANS, ROMA. *When Birds Change Their Feathers.* New York: Thomas Y. Crowell, 1980. Fascinating feather facts.

GOLDIN, AUGUSTA. *Ducks Don't Get Wet.* New York: Harper & Row, 1989. Good information, an experiment to explain the title concept, and fine prose in this revised classic. Paperback. Free activity card for teachers.

KAUFMAN, JOHN. *Birds Are Flying.* New York: Thomas Y. Crowell, 1979. Interesting information about how birds are able to fly.

PARSONS, ALEXANDRA. *Amazing Birds.* New York: Alfred A. Knopf, 1990. Outstanding color photographs and clear, easy text in this child's version of the London Natural History Museum Eyewitness Series. Paperback.

ROYSTON, ANGELA. *Birds.* Science Nature Guides. San Diego: Thunder Bay Press, 1994. Colorful illustrations and brief identifying characteristics of 150 common birds of North America from seven habitats. Birding activities are included.

RYDER, JOANNE. *Catching the Wind.* New York: Morrow, 1990. As a child soars in her imagination with the wild Canadian geese, readers absorb gentle lessons about the functional design of different kinds of feathers, beaks, and feet. AAAS recommended. Paperback.

SANTREY, LAURENCE. *Birds.* Mahwah, NJ: Troll, 1985. A brief, general reference covering common characteristics, nesting, and migration. Paperback.

SCHLEIN, MIRIAM. *Pigeons.* New York: Thomas Y. Crowell, 1989. These ubiquitous city birds are respectfully described by habits, rearing of their young, and historical lore.

YOLEN, JANE. *Honkers.* Boston: Little, Brown, 1993. Watching over three newly hatched goslings softens loneliness for Betsy, in this story of her summer away from home.

ZION, GENE. *No Roses for Harry.* New York: Harper, 1976. Nesting birds and Harry's detested sweater come together to solve problems. Paperback.

Stories and Resources About Mammals

CHERRY, LYNNE. *The Armadillo from Amarillo.* San Diego: Harcourt Brace, 1994. Written as a whimsically rhymed fantasy, much of the natural history and geographic information was researched on site under a Smithsonian Institution grant. The author's illustrations add rich detail and interesting perspectives to the curious armadillo's odyssey.

GOODALL, JANE. *Chimps.* New York: Byron Press Books, Simon & Schuster, 1989. Written to promote respect and understanding for animals and the habitat mankind must preserve for them, these Jane Goodall's Animal World books will appeal to older primary-grade children. Paperback. Other books in this series are: *Hippos, Lions,* and *Pandas.*

HOFFMAN, MARY. *Animals in the Wild: Elephant.* New York: Random House, 1985. Beautiful color photographs and simple information about young and mature animals in their natural surroundings. Paperback. Other books in this series are: *Animals in the Wild: Monkey, Animals in the Wild: Panda,* and *Animals in the Wild: Tiger.*

KAUFMAN, JOE. *Big Book About Mammals and Birds.* New York: Golden Books, 1989. Accurate information and cheerful illustrations are in abundance here. The front and back covers pose clever creative-thinking challenges.

LAVIES, BIANCA. *It's An Armadillo.* New York: Dutton, 1989. Armadillos can be fascinating and appealing when photographed at close range. Charmingly described by this patient photo journalist-author.

PARKER, STEVE. *Mammal.* New York: Alfred A. Knopf, 1989. Outstanding illustrations and well-organized information in this London Museum of Natural History Eyewitness Series book.

SABIN, FRANCENE. *Whales and Dolphins.* Mahwah, NJ: Troll, 1985. Presents corrective information about so-called "killer whales," the friendly, intelligent dolphin, and other seagoing mammals. Paperback.

STEIN, SARA. *Mouse*. San Diego: Harcourt Brace Jovanovich, 1985. This highly recommended book combines a warm, lean text with charming, accurate illustrations by Manuel Garcia. It reveals how mice nest, reproduce, and develop prenatally.

STUART, DEE. *The Astonishing Armadillo*. Minneapolis: Carolrhoda Books, 1993. This careful description of these armor-coated mammals includes fascinating photographs of newborns.

Stories and Resources About Habitats

CURRAN, EILEEN. *Life in the Forest*. Mahwah, NJ: Troll, 1985. Young preschoolers will enjoy locating forest animals hiding in nests and trees. Paperback.

CURRAN, EILEEN. *Life in the Pond*. Mahwah, NJ: Troll, 1985. A simple text for young preschoolers to enjoy, with vivid illustrations by Elizabeth Ellis. Paperback.

CURRAN, EILEEN. *Life in the Sea*. Mahwah, NJ: Troll, 1985. Beautiful and lively illustrations by Joel Snyder make this book special. Paperback.

EUGENE, TONI. *Creatures of the Woods*. Washington, DC: National Geographic Society, 1985. The color photographs forming the core of this book are intimate and dramatic. A simple text is supplemented with more detailed information for older children and adults.

HELLER, RUTH. *How to Hide a Butterfly*. New York: Platt & Munk, 1992. The clever, accurate rhyming text combines with the author's stunning illustrations to make an unforgettable book about protective camouflage in nature. Paperback.

HELLER, RUTH. *How to Hide a Whip-poor-will*. New York: Grossett & Dunlap, 1986. Stunning illustrations by the author reveal nature's protective coloration, body forms, and behaviors. Paperback.*

HOCKMAN, HILLARY (ED.). *Animal Homes*. New York: Dorling Kindersley, 1993. Vivid exterior photographs pair with cutaway drawings to peek inside a beehive, beaver lodge, squirrel nest, and rabbit warren, in this What's Inside Series book. Paperback.

KITCHEN, BERT. *And So They Build*. Cambridge, MA: Candlewick Press, 1993. A lovely text and striking, detailed paintings provide a respectful look at the intricate shelters built by animals.

MARTIN, JAMES. *Hiding Out*. New York: Crown Publishers, 1993. Remarkable closeup color photographs by Art Wolfe show ingenious animal camouflages.

RINARD, JUDITH. *The World Beneath Your Feet*. Washington, DC: National Geographic Society, 1985. Remarkable, delicate photographs, such as six ladybugs on a dandelion blossom, plus a simple text and supplementary information.*

RYDER, JOANNE. *Animals in the Wild*. New York: Golden Books, 1987. A nice presentation of seasonal activities of familiar woodland animals. Paperback.

RYDER, JOANNE. *Mockingbird Morning*. New York: Simon & Schuster, 1989. A nature-loving poet and an outstanding illustrator combine talents in this lovely child-view of animal life around her home.

RYDER, JOANNE. *Step Into the Night*. New York: Four Winds, 1988. The natural world at night is exquisitely portrayed in poetic text. A girl listens to, watches, and imaginatively becomes different nocturnal animals.

SABIN, FRANCENE. *Wonders of the Pond*. Mahwah, NJ: Troll, 1985. Explores animal life in the pond, from amoeba to muskrats. Paperback.

SABIN, LOUIS. *Wonders of the Desert*. Mahwah, NJ: Troll, 1985. Information about animal protection and survival mechanisms needed for desert life. Paperback.

SABIN, LOUIS. *Wonders of the Sea*. Mahwah, NJ: Troll, 1985. Describes sea animals from the microscopic to the gigantic, including the gentle, giant octopus. Paperback.

SELSAM, MILLICENT, & HUNT, JOYCE. *Keep Looking!* New York: Simon & Schuster, 1989. Beautiful in both words and watercolors, this fine book encourages patient observation of many forms of animal life in a winter backyard.

WYLER, ROSE. *Seashore Surprises*. Englewood Cliffs, NJ: Julian Messner, 1992. The fascinating habits of mollusks, crabs, and other shore life are revealed. Waves, sand, pebble, and saltwater activities are suggested. Nice watercolors capture the salty breezes.

ZOLOTOW, CHARLOTTE. *The Seashore Book.* New York: Harper Collins, 1993. Lovely textures in the illustrations echo the evocative descriptions of shore creatures seen during a child's imaginary day at the ocean. NASTA Outstanding Trade Book.

Stories and Resources About Baby Animals, Hatching, and Pets

BARBER, ANTONIA. *Gemma and the Baby Chick.* New York: Scholastic, 1993. A warmly told story of a girl who saves the almost ready-to-hatch chick, overlooked and abandoned by the brood hen.

BURTON, JANE. *Chick.* New York: Dutton, 1992. Bright photographs add understanding to a chick's personal narrative about its hatching and maturation. Also in this See How They Grow series for preschoolers are *Kitten, Puppy, Duck, Frog,* and *Rabbit.*

COHEN, MIRIAM. *Best Friends.* New York: Simon & Schuster, 1971. The friendship of Jim and Paul is cemented by their mutual rescue of the classroom incubation project.

COHEN, MIRIAM. *When Will I Read?* New York: Morrow, 1977. Danny's concern for the classroom hamsters helps him achieve his goal of becoming a reader.

DUGAN, BARBARA. *Leaving Home with a Pickle Jar.* New York: Greenwillow, 1993. Earnest gently transports his pet grasshopper in a pickle jar when he makes a cross-country move with his mother. Problems arise when Earnest allows his pet to exercise in the car.

GARLAND, SARAH. *Billy and Belle.* New York: Viking, 1993. While mother is in the hospital, Belle visits her big brother's school on pet day. She innocently manages to upset the whole menagerie while hunting for her lost pet spider.

HALL, DEREK. *Elephant Bathes.* New York: Alfred A. Knopf, 1985. Also in this series, *Gorilla Builds* and *Polar Bear Leaps.* Simple stories for preschoolers about young animals learning to cope with their natural environments.

HELLER, RUTH. *Chickens Aren't the Only Ones.* New York: Putnam, 1993. The simple and direct information in the text is vividly reinforced with the author's striking illustrations. A Children's Science Book Award winner. Paperback.*

INGOGLIA, GINA. *Nature Babies.* New York: Golden Books, 1989. The simple text in this small book offers young learners beginning information about habitat adaptation. Paperback.

KENAH, KATHERINE. *Eggs Over Easy.* New York: Dutton, 1993. This chapter book follows two city boys as they take responsibility for hatching the eggs they found in the park.

KNEIDEL, SALLY. *Pet Bugs: A Kid's Guide to Catching and Keeping Touchable Insects.* New York: Wiley, 1994. A wonderful resource for learning how to catch and keep 26 species of insects, with additional information about how they behave. Paperback.

MANUSHKIN, FRAN. *Puppies and Kittens.* New York: Golden Books, 1989. This small book about young pets nursing and developing characteristic behaviors is well-written for young learners. Paperback.

MCCLOSKEY, ROBERT. *Blueberries for Sal.* Seafarer Edition. New York: Viking Press, 1968. Four sets of mothers and their offspring are part of this story: Sal and her mother, plus bear, quail, and crow families.

MCCLOSKEY, ROBERT. *Make Way For Ducklings.* New York: Viking Press, 1969. Finding a suitable home and raising a brood of ducklings occupy Mrs. Mallard's time. Paperback.

RINARD, JUDITH. *Helping Our Animal Friends.* Washington, DC: National Geographic Society, 1985. Responsible care for pets by children, and professional care of orphaned wild animals.

STEIN, SARA. *Cat.* San Diego: Harcourt Brace Jovanovich, 1985. With simplicity and precision, the author focuses our attention on the predatory, self-sufficient nature of the family cat. The excellent text is enhanced by Manuel Garcia's clever illustrations.

STEIN, SARA. *Mouse.* San Diego: Harcourt Brace Jovanovich, 1985. This highly recommended book combines a warm, lean text with charming, accurate illustrations by Manuel Garcia. It reveals how mice reproduce and grow.

WARD, LYND. *The Biggest Bear.* New York: Scholastic, 1975. A touching lesson about the trials of trying to domesticate a woods animal.

WILKES, ANGELA. *My First Nature Book.* New York: Alfred A. Knopf, 1990. Simple bird feeding projects and directions for housing caterpillars and hatching moths and butterflies are shown in bright color photographs.

Stories and Resources About the Death of Pets

ROGERS, FRED. *Mister Rogers First Experience Book: When a Pet Dies.* New York: Putnam, 1988. Mister Rogers helps children manage their feelings of loss, loneliness, and frustration after a pet dies. Highly recommended.

VIORST, JUDITH. *The Tenth Good Thing About Barney.* New York: Atheneum, 1971. The conclusion of this story may not be acceptable to some. Decide for yourself.

Poems (Resources in Appendix A)

"The Little Land" from *A Child's Garden of Verses,* by Robert Louis Stevenson, records a charming perspective of insects projected by a diminutive sailor on a leaf boat. Many other poets create delicate or amusing sketches of insects and somewhat larger animals. Poems are found in these collections:

> de Regniers, Moore, and White, *Poems Children Will Sit Still for*
> Aileen Fisher, *When It Comes to Bugs*
> Paul Fleischman, *Joyful Noise: Poems for Two Voices.* Newberry Medal award poems evoke the delicate sounds of insects as they move.
> Dorothy Kennedy (Ed.), *I Thought I'd Take My Rat to School.* In X. J. Kennedy's poem, "Science Lesson," children tickle tadpoles.
> Pat Moon, *Earth Lines.* In "The Bird's Nest," children compete with a bird to build the best nest. They lose.
> Joanne Ryder, *Inside Turtle's Shell*
> Marilyn Singer, *Turtle in July*

Fingerplays

Many of the traditional fingerplays are about the characteristics of small animals. The following is an animal shelter fingerplay.

Here Is a Bunny	
Here is a bunny	(Fist with two fingers raised)
With ears so funny	
And here is his hole in the ground.	(Left hand on hip, arm curved)
When a noise he hears	(Stretch "ear" fingers)
pricks up his ears	
And jumps in his hole in the ground.	(Thrust fist through curved arm)

—AUTHOR UNKNOWN

Add to this fingerplay to include the movement of small animals observed by children in your classroom.

Small Animal Parade

A slow, slow snail

Drags down the trail.   (Stretch and contract right hand, dragging along left arm)

A looping earthworm

Moves along with a squirm.   (Loop and wiggle one finger)

A spider runs past

With eight legs so fast.   (Hands one on top of the other, tuck thumbs under, wiggle eight fingers)

The grasshopper springs

With six legs and wings   (One hand crossed on top of other, thumbs and little fingers held under. Leap up and down)

The green and white frog

Leaps over a log.   (Right fist leaps over left arm)

A seven-fin fish

Swims by with a swish.   (Palms together, hands twist and turn, moving forward)

—J.D.H.

Art Activities

Easel Painting. One of the early shapes many children paint spontaneously is a loop with many strokes radiating from it. Often these are called spiders or bugs by the painter. Provide green and brown paint on insect-watching day and casually suggest that some children could have fun making paintings of the spider, grasshopper, caterpillar, or whatever creature is of current interest. Newsprint sheets may also be cut into large butterflies or bird shapes for children to paint at the easel.

Crayon and Picture Collage. Provide pictures of animals (for children to cut out or have some precut for those new at cutting), paste, construction paper, and crayons for crayon-enhanced collages. Animal pictures are not easy to locate in the typical household magazines. Old children's magazines, free nature publications from your state natural resources department, or animal stamps from the National Wildlife Federation are good sources.

Leather and Feather Collage. Scraps of leather from a crafts shop and assorted feathers make interesting collage material. Use durable paper and white glue for best adhesion. (Feathers might be unacceptable collage materials in classes where allergic reactions to feathers are a problem.) Feathers can be collected from poultry farms, or on late-summer walks near lakes or wooded areas. Before use, wash feathers in this way: (1) Soak in cold water several hours; (2) wash in warm water and detergent, swishing and surging water through the feathers; (3) spread single layers on newspapers near a sunny window (no breeze, please). Let the children smooth the dry feathers with their fingers.

Additive Sculpture Animals. Pinecones are good bases for making animal forms. Turkeys can have rounded Scotch pinecone bodies, red painted balsam cone heads, and feather tails.

Use strips of molded paper egg cartons as caterpillar bodies. Offer tempera paint, fabric, paste, feathers, and snips of pipe cleaners for fanciful decorations.

Eggshell Mosaics. Broken, dyed eggshells can be glued to paper for egg-hatching project extensions. The younger the age group, the larger the shell bits should be for best management of materials.

Dramatic Play

Fishing. Add a dishpan full of water to indoor camping play. Tie a small magnet to a string. Attach to a stick. Cut simple fish shapes from pressed foam packaging sheets; attach a safety pin through it.

Animal Puppet Play. Put out animal hand puppets for children to animate. A small table turned on its side makes a good improvised stage.

Dramatize Animal Stories. When a group of children is familiar with an animal story that involves "a cast of thousands," they enjoy acting out the story in an informal, simplified way. Assign roles to all the children, so that no one is left out, even if someone has to play the part of a tree in the forest. Good animal stories that meet the casting requirements are: *The Story of Ping, Make Way For Ducklings*, and *The Tortoise and the Hare.* Help move the story along informally, as needed. "Maria, you're pretending to be Ping, so you stay back looking for food while all your family marches across the bridge." Assign areas in the room for different scenes. "Here's the island on the river where the Mallards build their nest, and the block area is the Public Garden where they swim." Use simple props if they help carry out the story. For example, arrange a few blocks in a boat outline to make Ping's houseboat for the duck relatives to crouch in.

Creative Movement

It's easy to draw children into the fun of acting out animal characteristics. Creative movement suggestions may also strengthen children's recall of the life cycles of animals that undergo metamorphosis. Slowly and dramatically read animal poems aloud for children to interpret with you. Good poems to try include the following found in *Poems To Grow On*, compiled by Jean McKee Thompson (see Appendix A):

"Cat," by Mary Britton Miller
"Mrs. Peck-Pigeon," by Eleanor Farjeon

"Baby Chick," by Aileen Fisher
"Fuzzy Wuzzy, Creepy, Crawly" by Lillian Schulz
"Earth Worm," by Mary McBurney Green

Clare Cherry suggests many animal-movement stimuli in her book, *Creative Movement for the Developing Child.* The Nature Company offers an imaginative sing-and-dance-along videotape, *Baboons, Butterflies and Me,* blending African wildlife footage with children imitating those movements. Appropriate for 2-to-6-year-olds (see Appendix A).

For an organized indoor play activity, children can be invited, one by one, to travel across the floor like the animal of his or her choice. (Groups of young children wait turns more patiently if they are sitting on the floor before and after their turn to travel.) One quick-thinking pair of 5-year-olds brightened our day when they piggybacked to crawl together as an 8-legged spider.

Creative Thinking

Habitat Classification. Pin pictures of a tree, lake, and grassy meadow to a flannel board. Mount small pictures, or cut out simple felt shapes, of an insect, fish, frog, turtle, snake, worm, spider, bird, and others. Let the children tell which picture "home" the creature belongs near. Can some animals live in more than one kind of "home"?

I'm Thinking of an Animal. Offer simple descriptions for children to guess, like: "I'm thinking of an animal that has no eyes, no legs, no arms; it has a mouth and a long body. What could it be?" Children who are experienced enough to take turns guessing will enjoy describing small mounted pictures drawn from a bag.

What If? "What if we had six legs to travel on? How would our lives be different? What could we do that we can't do with two legs? What couldn't we do? What if we had to spend most of our day gathering food for ourselves? What if we had working wings? How would our lives be different? What if we had to build places to live without using tools?" Encourage as many imaginative responses as possible. Let children generate their own "what if" questions to explore.

Animal Inventions. Encourage children to invent new animals, name them, draw what they would look like, and describe them.

Food Experiences

Children accept the idea that people eat fish and chicken, since the terms are customarily used at mealtimes. There seems little to be gained, however, from pressing the point that the hamburgers being served for lunch were once steer or that the ham in the casserole was once a pig. Concentrate instead on ideas that foster positive emotions: cows giving milk, hens giving eggs, or bees making honey for people to enjoy.

To confirm human dependence on other forms of animals for food, many cooking experiences are possible in the classroom: making butter (use fresh, not sterilized, whipping cream), instant puddings, ice cream, or custard.

Field Trips

The list of potential problems and hazards that might be part of an outdoor animal observation hike could rule out this kind of field trip for a large group of young children. It can also be very disappointing when the objects of the trip do not display themselves. Try instead to pause for watching time whenever you encounter small animals while outdoors with your children: ants at work in a sidewalk crack nest, limp worms washed out of their tunnels after a hard rain, or squirrels and birds gathering food.

Other field trip possibilities are to a farm, zoo, pet shop, or animal shelter. The state conservation department will have information about the location of animal preserves, hatcheries, or parks with programs by naturalists. They may also have free or inexpensive leaflets about animals.

Do not forget the grocery store as an animal study resource. Animals provide, indirectly, our many dairy products, eggs, and honey.

SECONDARY INTEGRATION

Maintaining Concepts

A teacher vividly reinforces concepts about the usefulness of insects by reacting calmly to an intruding bee or wasp. Remind children that insects of this sort sting only when disturbed. Open the windows from the top so that the insect can eventually fly out. Offer an observation jar to a shrieking child who is about to squash an uninvited spider. Do not undo what you have previously taught about the creature's place in the web of life by joining the chase with a can of insecticide. On the other hand, there is no need to be solicitous toward roaches, flies, or mosquitoes. If you must eliminate them in front of children, mention that they are pests to people but good food for birds and toads.

Another reinforcement of respect for living creatures may occur if a classroom pet dies or if a dead bird is found on the playground. Let the children be aware of the death and take part in a gentle burial. Say that the animal's useful life has ended. Try to read one of the stories about the death of animals that coincides with your views of death (see Stories and Resources About the Death of Pets earlier in this chapter).

Connecting Concepts

The broad ecological relationships between plants, animals, soil, water, air, and temperature are awesome and complex. Preschool and early elementary-age children can take beginning steps toward understanding ecological relationships through small, concrete instances. For example, a child may observe a bird pecking at tree bark and wonder aloud if the bird is hurting the tree. An adult can clarify the bird's immediate purpose, and expand the idea of mutual dependency. "Yes, it does look a bit like the bird is hurting the tree. Really, it is getting its food, and helping the tree at the same time. The bird is catching tiny insects that are eating the tree. Can you think of a way that the tree helps the bird?" Emphasize that plants and animals need each other.

A child may question the fitness of animals eating other animals, plants, and seeds for sustenance. One can appreciate the child's concern. Then suggest that each living thing can make many more seeds or eggs than are needed to make a young plant or animal like itself. If all the seeds grew into plants and all the eggs became animals, there would not be enough room in the world for all the living things. Some of the plants and animals need to be used in this way to keep the proper amount of each kind growing well to maintain the balance of nature. Young children can be upset by the recognition of our human role in the food chain. The fact that humans are animals that eat other animals may be better tolerated after the primary-grade years.

Sound concepts can be linked to the study of animal life. Compare the high pitch of the wren's song with the low pitch of the dove's coo. Feel the vibrations of a purring cat. Think of the swelling air pockets that make a frog's loud croak possible. Recall the vibrating wing covers that can be seen when the male cricket sings. Relate air concepts to the life requirements of animals. (See chapter 7, What Is Air?)

Family Involvement

Inform families early in the school year that children are encouraged to bring in captured insects. After the first few planned experiences you may never have to track down a specimen by yourself.

A knowledgeable parent might be willing to set up a classroom aquarium, make a pet cage, or arrange to bring in and show a family pet. His child's social standing with the group usually blossoms as a secondary benefit.

RESOURCES

BURNETT, ROBIN. *The Pillbug Project.* Washington, DC: National Science Teacher's Association, 1994. Provides science and math activities using these extremely accessible insects for observation. Paperback.

ECHOLS, JEAN. *Animal Defenses.* Berkeley, CA: Lawrence Hall of Science, 1990. Creative activity plans for exploring animal defenses.

ECHOLS, JEAN. *Buzzing a Hive.* Berkeley, CA: Lawrence Hall of Science, 1990. Art and drama activities for exploring behavior of bees.

Environmental Discovery Units. Inexpensive teaching guides for ecology study. National Wildlife Federation, 1412 16th St. N.W., Washington, DC 20036.

GABRIELSON, IRA, & ZIM, HERBERT. *Golden Guide to Birds.* New York: Golden Books, 1987. Pocket-size paperback.

GLASER, LAURA. *Wonderful Worms.* Brookfield, CT: Millbrook, 1992. Comprehensive, respectful treatment of this creature's ecological significance. NSTA Outstanding Science Trade Book for 1992.

HAMPTON, CAROL, AND KRAMER, DAVID. *Classroom Creature Culture: Algae to Anoles.* Washington, DC: National Science Teacher's Association, 1994. This collection of articles from *Science and Children* emphasizes respect for the needs of living things and is the resource to reach for when a new small creature joins the tabletop zoo. Paperback.

HARRISON, VIRGINIA. *The World of Honeybees.* Milwaukee: Gareth Stevens, 1990. Fascinating photographs and concepts provide a good background for helping a bee-phobic youngster in your class approach this insect safely.

KNEIDEL, SALLY. *Creepy Crawlies and the Scientific Method.* Golden, CO: Fulcrum Publishing, 1993. Guidance with finding, housing, and setting up experimental conditions for insects to respond to. The writing respects both the creatures and the learning children.

LEVI, HERBERT & LORNA. *Spiders and Their Kin.* New York: Golden Books, 1990. Pocket- size paperback.

PARKER, STEVE. *How Nature Works.* New York: Random House, 1992. Excellent background information on how animals meet their needs for food and air, and how they grow, move, and protect themselves. Detailed illustrations. Paperback.

RUSSELL, HELEN. *Ten-Minute Field Trips: Using the School Grounds for Environmental Studies* (2nd ed.), Washington, DC: National Science Teacher's Association, 1991.

STOKES, DONALD & LILLIAN. *The Bird Feeder Book.* Boston: Little, Brown, 1987. Paperback.

WILKES, ANGELA. *My First Nature Book.* New York: Alfred A. Knopf, 1990. Simple bird feeding projects. Directions for housing caterpillars and hatching moths and butterflies.

VIDEOS

Baboons, Butterflies and Me. 1992. The Nature Company, P.O. Box 2310, Berkeley, CA 94702.

Bugs Don't Bug Us. 1991. Bo Peep Productions, P.O. Box 982, Eureka, MT 59917. Insects in action in their natural settings, even changing into mature forms, plus creative movement and a good song to build courage about encountering these creatures.

The Human Body:

Care and Nourishment

Young children yearn to grow and become strong. They are eager to know what is inside their bodies. The experiences suggested in this chapter contribute to clarifying misconceptions and overcoming worries that children may have about their bodies. They help children take the beginning steps toward a lifelong responsibility for their own physical well-being. The following concepts are explored:

- Each person is unique.
- We learn through our senses.
- Bones help support our bodies.
- Muscles keep us moving, living, and breathing.
- We help ourselves stay healthy and grow strong.
- Strong, healthy bodies need nourishing food.

The goal of this chapter is to help children learn to value themselves as unique individuals, respect differences among individuals, and care for their own bodies. Mere knowledge of these concepts and health-care rules may not lead children into healthy practices. Positive attitudes about health lead to positive actions. Such attitudes grow through identification with a trusted person or symbol. With this in mind, a teacher might choose to introduce activities by using a puppet with whom young children can identify.

Within this chapter, specific experiences lead to ideas about individuality, the senses, bones, muscles, and the heart. Simple health care and nutrition concepts follow from these activities.

CONCEPT: Each person is unique.

1. What do I look like?

LEARNING OBJECTIVE: To gain understanding of bodily self-image.

MATERIALS:

Manila folder or construction-
   paper cover for each child

Paper punch, fasteners

Crayons

Paper

Yardstick or tape measure

Scale

Full-length mirror or several
   smaller mirrors

GETTING READY:

Prepare book covers; label

All About Me, (child's name).

Label page 1, "I am _____ tall,
   I weigh _____, on (date)."

Label page 2, "I look like this."

Fasten tape measure to wall
   securely, if used.

Play the tape "Everybody's Fancy," sung by Fred Rogers on *Won't You Be My Neighbor?*

SMALL-GROUP ACTIVITY:

1. Measure and weigh each child. Let child record the data on page 1. (Teasing about size differences hurts and must not be allowed.)
2. Encourage children to tell what they see in the mirror, then on page 2, draw what they see. Record each child's self-description on the page. (A withdrawn child may be unable to describe or draw himself. The teacher can help build the child's self-confidence by describing what the child's face looks like to her.)
3. Fasten completed pages in covers and tack to bulletin board until the next book entry is made.

2. Mine alone.

LEARNING OBJECTIVE: To appreciate one's uniqueness.

MATERIALS:

Two *All About Me* pages

Extra paper

Marker or crayon

Soft paper towels

Meat trays

Dry or premixed tempera
   paint, dark color

Water

Magnifying glasses

Scissors

Paste

GETTING READY:

Make paint pads for finger
   printing by folding damp
   towels to fit trays. Sprinkle
   dry tempera or pour thick,
   mixed tempera onto
   damp pad.

SMALL-GROUP ACTIVITY:

1. Invite children to use marker or crayon to trace around their hands, fingers spread wide, on the appropriately marked pages. (Offer help if needed.)
2. Compare hand tracings for similarities and differences. Suggest looking at palms and fingertips through magnifying glass. How do they look?
3. Show children how to lightly stamp fingertips on paint pads, then press their fingers firmly on extra paper. Encourage examining the prints with magnifiers to compare patterns of each fingertip. Discuss the uniqueness of prints and the use of fingerprints or footprints when babies are born in hospitals. Mention that the patterns children are printing are theirs alone.
4. After children have enjoyed making many prints, suggest making a careful print in the corresponding fingers of the hand tracings. Children with mature cutting skills may wish to cut out the completed hands to paste in their *All About Me* books.

Numbers are significant when they tell a boy how much he has grown.

Mark book pages "Mine Alone, Right" and "Mine Alone, Left." (Children's names should be put on pages promptly to avoid confusion.)

Handprints of the class members make a meaningful bulletin board display for a parent meeting, stressing that individuality is to be celebrated.

CONCEPT: We learn through our senses

Appreciation for the perceptual functions of our senses can be vividly experienced through their temporary absence. The following experiences make use of this principle.

1. We learn by hearing.

LEARNING OBJECTIVE: To appreciate hearing and to develop an understanding of people who are hearing impaired.

Group Experience: Introduce this experience by reading aloud *Mandy,* by Barbara Booth, to provide insights on hearing impairments. Learn what children already know about hearing aids.

Demonstrate what story time would be like for a child who could not hear by silently mouthing the words in a book while holding it up to share the pictures. If a television set is available, watch a program with the sound turned off. Encourage discussion about what it would be like to have a hearing loss. Would it be hard to learn and to play with friends at school? Without specific guidance, children find it hard to empathize with the limitations of a classmate whose hearing loss has no external signs.

Discuss how our ears function (see Chapter 14, "Sound"). Discuss ear safety hazards. Try to arrange a visit from the public health nurses who conduct audiology screenings or from a hearing impaired person who is willing to show children his hearing aid or demonstrate sign language. Talk about how much we learn about the world through our hearing. Take a listening walk to encourage children to appreciate their ears.

2. We learn by seeing.

LEARNING OBJECTIVE To appreciate vision and to develop an understanding of people who are visually impaired.

Introduction: Invite children to find out what it is like to recognize things and to move about without help from their eyes.

MATERIALS:

Lengths of soft paper toweling, long enough to fold into thick blindfolds for each child. (To avoid spreading eye infections, do not share blindfolds.)

Paper clips to secure blindfolds

Small objects to identify by touch. Mix easily identified items (brush, feather) with harder items (penny, nickel, 2 different books)

Small mirrors

Collection of objects to identify by sound: bell, clock, pieces of sandpaper to rub together, rubber band to snap

SMALL-GROUP ACTIVITY:

1. Fasten blindfolds. Since it is hard for children to stay blindfolded for more than a few minutes, pass only a few items from child to child for them to identify.
2. Loosen blindfolds. What did the children find out?
3. Fasten blindfolds. Encourage children to identify sounds. Remove blindfolds to let children use both eyes and ears to identify the sounds.
4. Let children examine their eyes in the mirrors and report what they see. Are the eyes still or moving? What do the lids do? How do their eyes feel when the children gently hold lids open without blinking for a minute? Is someone in the group wearing glasses? Why?
5. Offer to lead a blindfolded walk in the room. Fasten blindfolds for those willing to participate. Put children's hands on the shoulders of the child ahead. Slowly walk through the room, stopping to ask children where they are and how they know.

Read *A Cane in Her Hand* by Ada Litchfield. By trying cane traveling, children could gain empathy for people who are visually impaired.

3. We learn by smelling and tasting.

LEARNING OBJECTIVE: To recognize that some senses work together.

Introduction: Let children tell you what they already know about how food tastes when a head cold dulls the sense of smell. Suggest that children find out more about tasting and smelling at the science table.

MATERIALS:

Apple

Potato

Two bowls

Water

Lemon juice

Paring knife

Salt

Sugar

Gelatin dessert granules

Paper towel for each child

GETTING READY:

Peel apple and potato.

Slice thinly and drop into separate bowls of water.

Add a few drops of lemon juice to prevent darkening of the slices.

Read from *Smelling* and *Tasting* by Kathie Smith and Victoria Crenson.

SMALL-GROUP ACTIVITY:

1. Ask children to hold their noses closed. "Let's use our tongues to find out what we are eating. Taste these slices and decide what they are." (If the two foods look very different, have children take turns to help each other taste with eyes and nostrils closed.)
2. "Now taste your slices without holding your nose. How do they taste now?" Encourage more experimentation with tiny amounts of salt, sugar, and gelatin granules to find out if tongues alone tell us how food tastes.

4. What does our skin tell us?

Introduction: Ask the children to find out one thing about the contents of two containers that are on the science table without lifting or opening the containers.

MATERIALS:

Ice cubes or very cold water

Hot water

Two small containers with tight lids

Feathers

Chalk

GETTING READY:

Fill one container with ice water and the other with hot water.

Cover with layers of paper toweling to retain temperature differences until children can use them.

SMALL-GROUP ACTIVITY:

1. "What can you learn about what is inside the containers without opening them? How can you find out?" Which part of their bodies did they use? Could they find out the same thing by using their elbows? Foreheads? Backs of their hands? What covers each of these body parts?
2. Suggest finding out if it is easy for our skin to inform us about what it touches. "Try stroking your skin so gently with a feather that you can't feel it. Is it possible?"
3. Ask whether skin on some parts of the body gives better information about how things feel. "Does chalk feel the same when stroked across an arm; when rubbed by the fingertips?"
4. Encourage children to bring in an object with a special texture to put on a "Please Touch" tray.

Read from *Touching* by Kathie Smith and Victoria Crenson.

CONCEPT: **Bones help support our bodies.**

1. We have long bones and short bones.

LEARNING OBJECTIVE: To recognize that bone structure and body actions are related.

Introduction: Ask what children can feel deep inside their arms and legs when they gently press on them with a hand? Can they feel something in the middle of another child's back? Encourage them to investigate. The children could find out more about bones at the science table.

MATERIALS:

Chicken bones with rib section intact (mealtime left over)

Illustration of the human skeletal system

Desirable: Plastic model of the skeleton

GETTING READY:

Remove bits of breast meat from chicken rib section, leaving connective tissue and ribs intact.

SMALL-GROUP ACTIVITY:

1. Examine straight and curved chicken bones, long and short bones. Look at the connective tissue joining ribs. Help children feel their own ribs, tracing the curve of ribs from front to back. Are children's bones harder? Longer? (Bring out the idea that children grow bigger as their bones grow longer.)
2. Encourage children to curve their backs and curl up tight. Have them feel small bones in the spine and compare them with long bones in arms and legs. "Why do arms and legs bend differently than backs?" Have children feel joints that connect the long bones in arms and legs.
3. Encourage discussion of the bones children recognize in the model or illustration of the skeleton.

Read portions of *The Bones and Skeleton Book,* by Stephen Cumbaa, or *I Can Move,* by Mandy Suhr.

CONCEPT: **Muscles keep us moving, living, and breathing.**

1. Muscles keep us moving.

LEARNING OBJECTIVE: To identify the feeling of muscles in many parts of the human body.

Introduction: Begin the focus on muscles by guiding some seated stretches of the upper body. Let the children tell you what they already know about muscles. Point out that muscles move in two ways: by tightening, and by loosening. Add that it is easy for most people to use their muscles without even thinking about them. Other people have to think and work hard to use their muscles. They may need to have help from special braces, crutches, or chairs.

"We have hundreds of muscles in our bodies. Let's find out if we can feel some of them working to move our bodies."

Group Experience: This pattern for exploring the use and feeling of muscle groups is best done together, lying on the floor, but can be adapted to seated groups.

1. "Lift your fingers off the floor. Move just your fingers while the rest of your body is still and relaxed. Stretch your fingers high. Do you feel pulls in your hands and your wrists?"

2. "Now lift your whole hand off the floor. Lift just your hands while the rest of your body is still and relaxed. Turn your hands in circles. Do you feel pulls in your wrist and your arms?" Continue this way with lower arms and then with whole arms. Explore the pulls felt when the extended arm is slowly moved back and forth.

3. Continue the quiet, relaxed investigation of toe, foot, and ankle movements, asking children what they feel as they slowly lift, twist, and turn these body parts. When lying on their backs, can they lift their lower legs only? Why not?

4. Explore the muscles involved in curling up tightly, then slowly unfolding, sitting up, and twisting and turning the trunk.

5. Stretch the shoulder, neck, and jaw muscles. Appreciate smoothly working muscles. Conclude by using the facial muscles to make a smile.

Read *I Can Move* by Mandy Suhr.

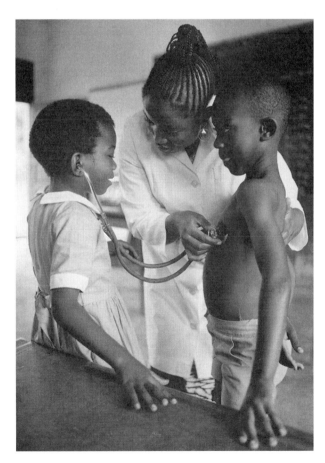

Robert's heart pumps, the visiting nurse facilitates, and Regina listens to life.

2. Your heart: A powerful muscle.

LEARNING OBJECTIVE: To identify the heart as a muscle.

Introduction: Ask the class to try for a moment to be perfectly still without moving a single muscle. Is it possible? No, fortunately! Can children feel some muscles continuously moving, even when their arms and legs are still? "Some muscles must continuously move to keep us breathing. We couldn't live if the mightiest muscle of all stopped moving; it is our blood-pumping muscle, the heart. The heart pushes blood through thin tubes to send energy and oxygen to all parts of the body. We can't see our heart working because it is safe inside our rib cage, but we can feel it working (demonstrate location)."

Show the children how to fold their hands together, then squeeze and release the grasp rhythmically. "Your heart works something like this, pumping and pushing day and night. Let's see how long we can do this with our hands. We can stop pushing our hands together when we get tired, but our hearts can never stop pumping."

MATERIALS:

Paper towel tubes

16" (40 cm) lengths of garden hose

Watch with second hand

Stethoscope (desirable)

SMALL-GROUP ACTIVITY:

1. Ask children to fold their clasped hands next to their ears to listen as they squeeze their hands together. "Your heart makes a soft thumping sound like that each time it pumps and relaxes."
2. Children can listen through paper tubes to hear one another's heartbeats. One ear rests on the tube end, the other is covered with a hand. Children can curve the hose to listen to their own beating hearts. (If a stethoscope is used, wipe earpieces with alcohol after each child's use.)
3. Try counting heartbeats for half a minute when children are quiet. Compare with heart rate after children jump in place for half a minute. What happens during exercise?

Read from *Your Insides* by Joanna Cole.

Note: If a child seems worried about the increase in heart rate during exercise, reassure the child with more information: "Heart muscles grow stronger and work even better when they are well-exercised." Children may have much to contribute to a discussion about muscle-building exercises.

3. What pushes our lungs?

LEARNING OBJECTIVE: To recognize that muscles keep us breathing.

Group Experience: Discuss children's recollections about changes in the heart rate after exercising. Ask whether they have noticed another change in their bodies when they run hard. Do they ever pant after exercising hard? "Our breathing gets faster, just as the heart rate gets faster. Breathing and heart rates change together to meet the changing needs of the body. Hard-working muscles need a fast supply of blood and oxygen to get energy. The heart works hard to pump blood faster and other muscles speed up to pull oxygen from the air into the lungs." In the following activity, children can find out more about how these muscles lift the rib cage to let new air come into the lungs.

MATERIALS:

Yardstick (or meterstick)

Large mirror

Card or stiff paper for each child

Skeleton model or illustration

SMALL-GROUP ACTIVITY:

1. Look at the curved rib cage on the skeleton model or illustration. The rib cage protects important parts inside our bodies like the heart and lungs.
2. Have each child find his or her lower rib, then hold a card at right angles to the rib. What can be seen in the mirror? Is the card moving? How? Hold the yardstick next to a child as he or she watches the card rising and falling. How far does the card move? Special muscles near our midsections move our rib cages up and down. Why?

Read *How I Breathe* by Mandy Suhr.

Note: A highly recommended guide for further meaningful activities to understand the body systems is described in Helen H. Johnson's article, "The Bodyworks: Inside Me. Another Approach to Alike and Different." (See Resources, p. 128.)

CONCEPT: **We help ourselves stay healthy and grow strong.**

1. *Our bodies need rest and exercise.*

LEARNING OBJECTIVE: To recognize that rest is essential to ensure a healthy body.

Introduction: Our bodies get tired from working hard: When we are very active, our muscles can use up the nourishment and oxygen they get from the bloodstream. Then the muscles lack energy to keep working. They are tired. They need to relax and rest before they can go on working hard. Other things happen to all of us when we are tired at the end of the day. Our eyes get sleepy, our minds don't think as well, or we may feel cross. We need rest. Children have another important reason for slowing down to rest. *Their bodies need extra energy to grow.* When they rest and sleep at night (about 10 hours), their bodies can get ready for another day of working, playing, and growing. Children can draw or cut out and paste pictures of these two things their bodies need for health and growth: Exercise and rest.

MATERIALS:

Scissors

Paste

Crayons

Magazine or catalog pictures of active and resting people

All About Me pages

SMALL-GROUP ACTIVITY:

1. Encourage children to draw themselves enjoying exercise, or to cut and paste pictures that remind them to exercise and rest.
2. Help children write a realistic statement about how much sleep they need to stay healthy.

Read *Why Do We Need Sleep?* by Isaac Asimov and Carrie Dierks.

2. *Soap washes away dirt.*

LEARNING OBJECTIVE: To observe the cleansing effect of soap and water.

Introduction: To spur interest in the topic of handwashing, share some surprising informa-tion with the class: "One of the greatest disease-preventing inventions of all time is the bar of soap! The soap we use to wash our hands helps loosen and wash away illness-causing germs on our skin." Find out what children know about germs. Supplement their responses, if necessary, with the fact that the germs that cause illness are too small to see. Unless soap helps to wash them off, germs may still be on the hands when children eat. Then the germs can spread onto their food. Children could experiment to find out more about the effects of washing with soap and water.

MATERIALS:

Old white fabric

Two dishpans

Bar of soap

Water

Paper towels

Two trays

Paper and paste

Scissors

Potting soil

Oil

GETTING READY:

Experiment with water, soil,
  and small amounts of oil
  to make thin solution.

Cut off pieces of the fabric
  to save as control samples.

Dip remaining fabric into
  soil solution. Let dry and
  cut into pieces, allowing
  two for each child.

Collect pictures of soap and
  of people washing.

Mark *All About Me* pages: "I
  wash myself with soap."

SMALL-GROUP ACTIVITY:

1. Let children wash one sample of dirty fabric in a dishpan of water and a second sample of dirty fabric in a separate dishpan of water, scrubbing it with a bar of soap.
2. Line two trays with paper towels. Put the water-washed samples on tray 1 and the soap-and-water-washed samples on tray 2. Mark children's initials next to damp samples.
3. Compare dried fabric samples to see which ones are cleaner. Paste samples on booklet page next to a sample of the control fabric for comparison. Label.

3. Immunization helps us stay well.

LEARNING OBJECTIVE: To develop acceptance of the need for immunization.

Large-Group Experience: Encourage children to share what they know about measles, mumps, and other childhood illnesses caused by viruses. Read aloud selectively from *Germs Make Me Sick!* by Melvin Berger. Place special emphasis on page 27, which deals with giv-ing shots to prevent viral diseases that can't be cured with drugs. Support children who ex-press fear of getting shots. Additional reading about shots from *Mister Rogers Talks About Going to the Doctor* may be helpful to younger children.

Supplement the reading with an opportunity for children to examine a liquid medi-cine dispensing syringe from the drugstore. Let them practice filling it with water and emptying it. Set up a hospital play situation, including the syringe, for preschoolers. Invite a public health nurse to visit to provide information about immunization in a nonthreatening way.

Add a child-oriented message on immunization to the *All About Me* booklets. Local offices of national health groups, or the public health office in your community, may have immunization promotional leaflets suitable for inclusion in the booklets. Immunization record booklets from the public health office could be sent home to parents to reinforce the classroom focus.

4. We take care of our teeth.

LEARNING OBJECTIVE: To recognize responsibility for one's own dental health.

Introduction: Read aloud Paul Showers' humorous book, *How Many Teeth?* Give children the opportunity to share their ideas about loosening and erupting teeth.

Invite a dental hygienist to present introductory dental health care information to the children. Some professionals are prepared to demonstrate good tooth-brushing techniques with sets of plastic teeth and giant toothbrushes. If this is not possible, ask children to share what they already know and do to take good care of their teeth. Supplement their comments, as needed, to include regular dental checkups and eating the foods that make strong teeth and healthy gums, especially milk-group foods. Talk about smart snacks for healthy teeth.

5. What do our teeth do?

MATERIALS:

Small mirrors

Tissues

Graham crackers

Slices of apple

Magazine pictures of tooth-paste, toothbrushes, and children brushing their teeth

All About Me booklet pages

Paste

Scissors

GETTING READY:

Mark booklet pages "I Brush My Teeth."

Fold a tissue into a small pad for each child.

SMALL-GROUP ACTIVITY:

1. Encourage children to smile at themselves in the mirror. What does the smile look like? Are teeth useful for more than smiles? Are teeth all alike?
2. Let children carefully bite down on the tissue pads with all their teeth. "Look at the marks on the pads. Are they all alike? Which teeth would be best for biting? Which teeth would be best for chewing and grinding food fine enough to swallow?"
3. Encourage children to use mirrors to examine the grinding teeth. Give children bits of cracker to chew. Use the mirrors to view the grinding surfaces again. Can the tongue clean away some cracker crumbs?
4. Offer apple slices to the children to chew thoroughly. Check with mirrors again. Has the apple, "nature's toothbrush," changed the way the grinding teeth look?
5. Children may draw or paste magazine illustrations of tooth brushing on booklet pages.
6. Join the children at the sink to practice swishing water around in the mouth to rinse away food bits.

FIGURE 6–1   Food Guide Pyramid

CONCEPT: **Strong, growing bodies need nourishing food.**

1. What is a healthy, balanced diet?

LEARNING OBJECTIVE: To recognize and appreciate nourishing foods.

Large-Group Experience: Read, illustrate, and discuss the following flannel board story: "The Boy Who Wanted to Eat Upside Down."

Getting Ready: To illustrate the story, form a yarn outline of the Food Group Pyramid on the flannel board: Make an equilateral triangle, divided into four levels. Use smaller pieces of yarn to divide the two middle levels into appropriate sections. Cut out and mount pictures of a variety of foods on adhesive backing. Place pictures in appropriate food group sections as you read the story. Or, use an attractive Food Guide Pyramid poster, pointing to the foods as the story progresses. (See Figure 6–1.)

THE BOY WHO WANTED TO EAT UPSIDE DOWN

A boy was curled up in a comfy chair with his book when his grandma came back from the grocery store. She put down the bags full of groceries and pulled out a box of graham crackers. She said, "I have something important to show you. See this picture on the box? It's called the Food Guide Pyramid. It shows us how to eat for good health. From now on, we're going to follow this guide to healthy eating every day."

The boy looked at the triangle picture on the box. He saw that it had four levels. Two levels had two parts. Each part had food pictures in it. The biggest level was at the bottom. Grandma pointed to it. "The biggest part is called the Bread Group. A child your size needs to eat 6 servings of this kind of food every day, so I bought bread and cereal, rice and macaroni for you." The boy said nothing.

"Above the bread level is the vegetable and fruit level." Grandma went on. "I bought carrots and broccoli and sweet potatoes from the Vegetable Group, and apples and oranges from the Fruit Group for you. I'm going to give you 3 servings of vegetables and 2 servings of fruit every day." Still the boy said nothing.

Next Grandma pointed to two groups in the protein and minerals level. "That's the Milk Group. A child your size needs 3 servings every day, so I bought milk and cheese and yummy yogurt for you. And this is the Meat Group. A child your size needs 2 servings every day, so I bought some chicken and some beans to make chili for you." The boy said nothing.

Grandma said, "Now you can see how the picture helps us figure out what you need to eat the most of, and what you need a bit less of, to be healthy, strong, and growing. You need to eat the most servings from the largest level at the bottom of the pyramid, and smaller amounts from the levels closer to the top. See what's at the very top in the smallest level? Just two kinds of speckles: They stand for fat and sugar. Foods like cookies and cake and chips and candy and sugary sodas are in that level. We don't need much of those foods at all. We'll just have them once in a while, and not too much then."

Then the boy spoke. "If we get to eat the most of what's at the bottom level, then this is the way I'm going to be." He scooted around so that his head hung down from the comfy chair. "I wish I could be a bat, or a three-toed sloth, or a nuthatch bird so I could eat upside down, like they do. You can just serve me all the sweets and desserts and chips and fizzy drinks that I want. When I eat upside down, they're on the bottom of the pyramid! That's what I want to eat every day for the rest of my life."

"Of all things!" Grandma said. "We can't have that wish coming true! It's a good thing you're not a bat or a nuthatch or a three-toed sloth hanging about upside down all day, just eating sweets and chips and buttery snacks and sugary drinks. If you didn't eat enough foods from the protein level, your muscles and bones wouldn't be strong enough to hang on to your perch upside down. Your fur or feathers would look terrible. Your teeth wouldn't be good enough to chew your fatty, sugary foods, and you wouldn't grow strong bones."

"Umm," said the boy. "I wouldn't like that!"

"If you didn't eat enough from the fruits and vegetables level, you wouldn't have the vitamins to help your body stay well," Grandma said. "You'd be hanging there half sick, coughing and sneezing and feeling miserable all the time."

"Oh," said the boy. "I wouldn't like that!"

"If you didn't eat bread and cereals like oats and wheat and rice, and pasta like spaghetti and macaroni, you wouldn't have the energy you need to move around up there. You would just be tired and droopy. Does that sound like much fun?" Grandma asked.

"No," said the boy. "I wouldn't like that at all!"

"And you need to know something else," Grandma added. "If you only ate desserts and crispy chips and fizzy drinks, you would grow so pudgy and heavy that you would finally collapse and fall from your perch."

The boy began to laugh and giggle so hard that he nearly slid right down off the chair in a heap. Just in time he caught hold of the chair and pulled himself rightside up. He said, "OK, Grandma. I guess I don't want to be a bat, or a nuthatch, or a three-toed sloth. I'm going to have what you bought for me, rightside up. When do we eat?"

2. How does each food group help us?

LEARNING OBJECTIVE: To recognize the health contribution of different foods.

Group Experience: Play this food group classification game.

MATERIALS:

Six grocery bags: 1 large-sized, 4
  medium-sized, 1 small-size

Food replicas; mounted food pictures; empty food cartons and containers to represent all food groups

Food Pyramid poster or model

GETTING READY:

Label largest bag: BREADS AND
  GRAINS (6)

Label 4 medium-sized bags:
  VEGETABLE GROUP (3)
  FRUIT GROUP (2)
  MILK GROUP (3)
  MEAT GROUP (2)

Label smallest bag:
  FATS AND SWEETS

LARGE-GROUP ACTIVITY:

1. Ask: "If we brought groceries home from
   the store in separate food group bags, how
   would we do it?"
2. Invite each child to choose a food carton or
   picture to place in a grocery bag. As the item
   is deposited in the appropriate bag, mention
   the daily nutritional need the group provides.

Breads and Grains Group:	"These provide energy and fiber. Cereals and pasta be-long here, too. We need to eat the most from this group. Children need 6 servings each day."
Vegetable Group:	"Vegetables have lots of vitamins. Vitamins help good foods work to keep our bodies healthy and strong. Children need 3 servings each day."
Fruit Group:	"Each kind of fruit gives us special vitamins to stay healthy. Children need at least 2 servings each day."
Milk Group:	"Milk, cheese, and yogurt help build strong teeth and bones. Growing children need 3 servings each day."
Meat Group:	"Fish, eggs, nuts, and all kinds of dried beans belong here. They help bodies grow and develop strong mus-cles. Children need 2 servings every day."
Pyramid Tip:	"Foods that have lots of fat or oil or sugar belong here. They are not a needed food group. 'Tip' foods don't help us grow well and strong, nor stay healthy. We should only eat a little bit of these foods."

Note: Give this activity plenty of time. Empty the bags and repeat, so that each child has several chances to classify foods.

3. How are grains used as food?

LEARNING OBJECTIVE: To appreciate the many uses of nourishing whole grains.

The Food Pyramid guidelines emphasize the importance of grain foods as the basis of healthy, balanced diets. Wheatberries (hulled wheat) are low in cost and highly nutritious. This is an interesting whole grain to explore in the classroom. Now returning into vogue, cooked wheatberries, such as a dish called frumenty, was introduced to this country by colonists and was a staple in pioneer diets. These food experiences link well with studies of early settlers. Whole grains can be purchased in natural food stores, or ordered by mail. (See p. 129.) If a child in your class is allergic to wheat, whole hulled barley can be substituted. It can be ground into flour as with wheat. For barley cooking and sprouting directions, consult Jane Brody's *Good Food Book.* (See Resources, p. 128.) One pound (454 grams) of whole wheatberries will suffice for all of the following three whole grain activities. If possible, display a small bunch of dried wheat sheaves in the room.

Part 1: Cooking Wheat

MATERIALS:

Whole wheatberries

Measuring cup

Clean tray

Sieve

Water

2-quart saucepan

Salt

Electric crockpot

Alternative equipment, if classroom cooking is not possible:

Widemouthed quart thermos jug

Teakettle

Facility to boil water

Large-Group Experience: Show hulled, whole wheat grains and wheat sheaves to class. Let them feel the hardness of the wheatberries. Invite children to share their knowledge of wheat as a familiar food. Near the end of the school day, have children measure 1 cup (200 ml) wheatberries, then spread them out on the tray to sort out and remove broken grains and chaff. Save some grains in a small container for later comparisons with cooked, ground, and sprouted products.

Carefully transfer sorted wheatberries to the sieve, rinse under running water, put them in the pan, and cover with 2 cups (400 ml) of water to soak overnight. The next morning, let children watch as you drain off the soaking water, put the soaked wheatberries in the crockpot, and add enough warm water to cover the berries. Stir in 1/2 teaspoon salt. Cover and cook at highest setting for 2-1/2 to 3 hours. Grains will have doubled in bulk and be tender. They will be somewhat chewier than corn niblets. Most of the water will be absorbed. Serve samples to each child.

Safety Precautions: Locate the crockpot and the cord in a safe place, blocked from the reach of children. Remove the hot cover carefully, away from children. Unplug the cord when the cooking period ends. Move the pot away from the electric outlet.

Alternative preparation method: Take the soaked berries home or to the school kitchen, away from children. Then:

1. Boil kettleful of water. Rinse the thermos bottle with some of the boiling water. Heat thermos by filling it completely with boiling water. Cover to keep warm. Empty the hot rinse water just before adding the heated wheat in the next step.

2. Heat wheat and soaking water in saucepan. Bring to a vigorous boil. Add salt.

3. Fill empty, warmed thermos with hot wheat and water. Add a little boiling water if needed to just cover the wheat. Leave some air space in the neck of the bottle. Screw on stopper and cap securely. Place bottle on its side. Leave for at least 8 hours. Soaked wheat placed in the thermos bottle in the evening will be ready to eat in the morning when you take it back to school.

Part 2: Grinding Wheat into Flour

This can be a satisfying activity when children can take turns using an old-fashioned coffee grinder, if one can be found.* They will still be impressed by the grinding process, even if they can only watch their teacher do the grinding with an electric food blender. (Do not use a food processor. It would be damaged by the tough kernels.) Place 1 cup of wheatberries in a glass (not plastic) blender container, and process at high speed for about 4 minutes, until all the wheat is ground. Sift the flour through the fine sieve into a bowl. The bits of grain that don't go through the sieve are called grits. They can be cooked as breakfast cereal.

Safety Precaution: Keep one hand on the blender throughout this period. Do not allow children to touch the blender. Unplug from the electric outlet when grinding is finished.

Mention that for thousands of years before newer grinders were invented, millers ground grains between heavy, grooved millstones to make flour. If possible, bake something at home with the whole wheat flour. Share it with the class.

Read aloud "Long Winter Bread" from *The Little House Cookbook: Frontier Foods from Laura Ingalls Wilder's Classic Stories*, by Barbara Walker. (See Resources, p. 129.)

Part 3: Sprouting Wheat

Sprout wheatberries, following the directions in the activity that follows (Sprouting a Vitamin Crop). Wheat sprouts will be ready to eat in 2 days. (Sprouted wheat tastes quite sweet.) Mention that seeds develop vitamins as they begin to sprout.

In a summarizing group discussion, pass the small container of reserved wheatberries. Talk about the three ways wheat seeds were changed so they could be eaten. Encourage children to look for pictures of wheat or other grains in magazine ads, or on cereal, pasta, and cracker boxes when they grocery shop with their families.

4. Sprouting a vitamin crop

LEARNING OBJECTIVE: To experience growing and eating nutritious food.

Seed sprouting is probably the least expensive nutrition activity for the classroom. A few cents worth of seeds will produce enough sprouts to make an appealing snack for the whole class. The sprouts are rich in vitamins and minerals. The sprouting process can involve each child during the plant germination period.

MATERIALS:

Alfalfa seeds from a natural foods store

Small-mouthed quart jar

6-ounce foam cup

Thick needle

5" (12.5 cm) square of porous fabric

Rubber band

Cellophane tape, paper

Saucer

SMALL-GROUP ACTIVITY:

1. Discuss the sprouting project with the children on a Friday. Tape a few dry alfalfa seeds to a piece of paper to use for a day-by-day comparison with the changing sprouts.
2. Let children measure 2 tablespoons of alfalfa seeds into the jar. Explain that you will add one cup of warm water to cover the seeds in the jar at your home on Sunday night.

---

*Alternatively, a small electric coffee grinder can be used **by the teacher** to grind 1/4 cup of wheatberries for 4 minutes. **Safety Precaution:** Unplug this appliance immediately after use. Always keep out of the reach of children.

GETTING READY:

Method 1

Use a needle to pierce the bottom and lower half of the cup generously to form drain holes. (Figure 6–2.)

If class is large, prepare two sprouting jars to allow more child participation and to provide more sprouts.

Method 2

Follow the same procedures as for Method 1, except use a 2-liter plastic beverage bottle for a sprouting container.

Cut off the top quarter of the bottle and slide a nylon knee sock over the open end. Knot the top of the sock as shown in Figure 6–3.

Rinse and drain sprouts through the nylon mesh. Hold by knotted end to drain.

3. On Monday morning, demonstrate how to cover the jar top with the fabric, then fasten it with a rubber band so the water can be carefully drained off. Refill the jar with cool water and let children gently swirl the seeds in the fresh water. Then carefully drain the water.

4. For Method 1, remove fabric and insert drain cup into the mouth of the jar as far as it will fit. Invert the jar so the cup rim serves as a stand, as shown in Figure 6–2. Place on a saucer. Keep it away from direct sunlight.

5. Set up a rinsing/draining schedule two or three times each day so that each child has a chance to care for the sprouting seeds. As the sprouts begin to develop, the foam cup can be used instead of fabric for draining and rinsing the sprouts in Method 1.

6. As tiny green leaves appear, the jar may be kept in sunlight to allow the sprouts to develop chlorophyll.

7. On Friday, empty sprouts into a deep bowl. Cover with fresh water. Any hard seeds that failed to sprout will sink to the bottom of the bowl. Scoop out only the sprouted seeds near the top. Drain on toweling. Serve to children.

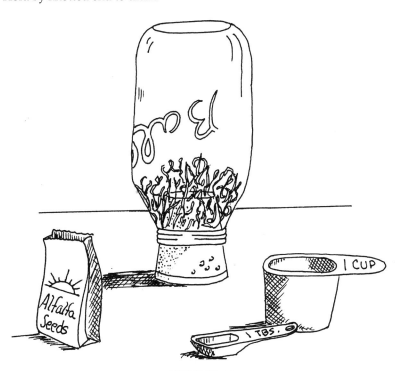

FIGURE 6–2

FIGURE 6–3

PRIMARY INTEGRATING ACTIVITIES

Math Experiences

1. Translate into concrete experience the evidence of growth from a child's birth weight to present weight. Do this for each child by stacking enough hardwood blocks on a bathroom scale to equal the two weights. Children may want to enter the following information in their *All About Me* books: When I was a baby, I weighed as much as _____ blocks; now I weigh as much as _____ blocks. Each child can feel good about her or his present state of growth when it is compared with past growth.

2. Establish a relative measure of muscle strength by weighing the number of blocks a child can lift. (Avoid competitive emphasis by using questions like, "How much weight are your muscles ready to lift now?")

3. To develop a sense of how long a minute is, let children use an egg timer or watch the second hand of a clock to time various enjoyable aerobic exercises. Have children count their heartbeats for 10 seconds before and after vigorous exercise.

4. Have children use their hands and feet to measure distances or dimensions: "This table is as wide as 11 of my hands."

5. Use fruit and vegetable shapes for flannel board math activities in set formation, one-to-one correspondence, and patterning.

6. Allow time for children to compare measuring spoons and cup sizes when they are used in cooking experiences. Try to have enough measuring equipment on hand so that all the children can participate in measuring ingredients. "We need one cup of flour for the dough. Will we have enough if two children each put in one-half cup of flour? Let's find out." Use sets of plastic measuring cups at the sand and water tables, so that children can practice using these utensils and develop a practical sense of their relative volumes.

Music (Resources in Appendix 4)

1. Raffi's *Singable Songbook* offers "I Wonder If I'm Growing?" and "Brush Your Teeth" to extend human body study. "Popcorn" and "Peanut Butter Sandwich" extend nutrition concepts.
2. The *Second Raffi Songbook* offers "All I Really Need," "Oats, Beans, and Barley," and "Biscuits in the Oven" to extend nutrition concepts.
3. Listen to "You are Special" from Mister Rogers' cassette *Won't You Be My Neighbor?* and "Everybody's Fancy" and "Everything Grows Together," from *Let's Be Together Today.*
4. Listen to how our bodies work in "What A Miracle," from Hap Palmer's cassette, *Walter the Waltzing Worm.*
5. "Here We Go 'Round the Mulberry Bush" provides a good pattern for improvising a song about health care. Children can sing, "This is the way I take care of myself . . . so I'll be strong and healthy." Add verses such as "This is the way I (brush my teeth, take my bath, wash my hands, go to sleep, eat good meals) . . . so I'll be strong and healthy."

Stories and Resources About Individuality

BOOTH, BARBARA. *Mandy.* New York: Lothrop, Lee & Shepard, 1991. This gentle story provides insights into how a profoundly hearing impaired child compensates for her disability with other senses, lip-reading skill, and courage to solve a problem.

CARLSON, NANCY. *Arnie and the New Kid.* New York: Viking Penguin, 1990. Classmates and a wheelchair-bound child learn to accept their differences and enjoy their similarities.

CASELEY, JUDITH. *Harry and Willy and Carrothead.* New York: Greenwillow, 1991. A child with a prosthesis teaches others about acceptance and understanding.

KEATS, EZRA JACK. *Apt. 3.* New York: Simon & Schuster, 1986. Sam learns to appreciate how his new sightless friend learns about his world. Paperback.

LITCHFIELD, ADA. *A Button in Her Ear.* Chicago: Albert Whitman, 1976. The story follows the detection and correction of a child's hearing loss. Angela's teacher helps build acceptance among classmates for Angela's hearing aid and battery harness.

LITCHFIELD, ADA. *A Cane in Her Hand.* Chicago: Albert Whitman, 1977. Val encounters difficulties when her vision problem develops. She learns how to use a cane "like a long arm" to help her adapt to her limitation and carry on with activities she enjoys.

LITCHFIELD, ADA. *Captain Hook, That's Me.* New York: Walker & Co., 1982. A girl born with one hand copes with her limitation. For primary-grade children.

MULDOON, KATHLEEN. *Princess Pooh.* Chicago, IL: Albert Whitman, 1989. A girl learns about the limitations experienced by her wheelchair-confined sister.

RUSSO, MARISABINA. *Alex Is My Friend.* New York: Greenwillow, 1992. Alex has serious growth problems. His humor helps him cope with grim orthopedic surgery and a wheelchair-bound future.

STERLING, BARBARA. *I'm Not So Different.* New York: Golden Books, 1986. Kit's story is designed to win respect and acceptance from children whose legs do for them what Kit's wheelchair does for her.

Stories and Resources About the Body

ALIKI. *I'm Growing!* New York: Harper Collins, 1992. It's all here: simple information about growth relevant to young children, presented in Aliki's clear, enjoyable style.

ALIKI. *My Five Senses.* New York: Harper & Row, 1989. A charming child enjoys experiencing the world through sensory awareness. Preschool. Paperback.

ANDERSON, KAREN, & CUMBAA, STEPHEN. *The Bones & Skeleton Game Book.* New York: Workman Publishing, 1993. Interesting, simple experiments will extend children's understanding and appreciation of human body intricacies.

ARDLEY, NEIL. *The Science Book of the Senses.* San Diego: Harcourt Brace Jovanovich, 1992. This attractive book of sensory experiments provides fresh facts and enrichment activities.

BALESTRINO, PHILLIP. *The Skeleton Inside You.* New York: Harper & Row, 1989. Describes the supportive and protective functions of our bone structure. A good classroom reference. Paperback.[*]

BRANDT, KEITH. *The Five Senses.* Mahwah, NJ: Troll, 1985. A good narrative style presents accurate information about senses as part of the nervous system. Paperback.[*]

COLE, JOANNA. *The Magic Schoolbus: Inside the Human Body.* New York: Scholastic, 1989. This clever fantasy trip down the esophagus into the stomach, bloodstream, and major organs shows how the body converts food to energy for body cells to use. Highly recommended to draw children into respecting their bodies! Paperback.[*]

COLE, JOANNA. *Your Insides.* New York: Putnam & Grosset, 1992. A wonderful book that places information about body systems into meaningful child context. Cheerful illustrations by Paul Meisel use young child body proportions. Acrylic overlay pages offer an inside view of the location of organs and systems.

CUMBAA, STEPHEN. *The Bones and Skeleton Book.* New York: Workman Publishing, 1990. Sold as a companion guidebook to an excellent ready-to-assemble plastic skeleton model, this book also adds interesting child-level information on other body systems.

HOBAN, LILLIAN. *Arthur's Loose Tooth.* New York: Harper & Row, 1985. Arthur copes with the experience of losing his first tooth.

KAUFMAN, JOE. *Joe Kaufman's Big Book About the Human Body.* New York: Golden Books, 1987. Excellent beginning information about the structures of the body, laced with the author's lighthearted illustrations. The accurate information provides valuable correction for the common misconception that our lungs are like balloons.

PARKER, STEVE. *Human Body.* New York: Dorling Kindersley, 1994. This Eyewitness Explorers book offers brief, well-chosen information about how our bodies work. Some simple experiments include a demonstration of why we need to wash our hands.[*]

ROJANY, LISA. *Exploring the Human Body.* Hauppage, NY: Barron's Educational Series, 1992. This book strengthens its clear text with pull tabs to move joints, flaps to lift, and transparent pages to view inner structures and systems. It also includes activities and experiments for this age group.

SABIN, FRANCENE. *Human Body.* Mahwah, NJ: Troll, 1985. A good reference on anatomy and physiology for any early childhood classroom. Simple, clear illustrations by Don Sibley. Paperback.[*]

SHOWERS, PAUL. *Ears Are For Hearing.* New York: Thomas Y. Crowell, 1990. Older children will find this careful explanation of the inner ear workings and causes of hearing loss interesting. Child-level analogies make the topic understandable. Paperback.[*]

SHOWERS, PAUL. *How Many Teeth?* New York: Harper Collins, 1991. The amusing text and illustrations enliven the subjects of deciduous teeth, good dental care, and a lisping story demonstrating how teeth help us speak. Paperback.

SHOWERS, PAUL. *Your Skin and Mine.* New York: Harper Collins, 1991. This book presents a simple treatment of the structure of the skin, the healing of scrapes, and the function of melanin. Paperback.

SMITH, KATHIE, & CRENSON, VICTORIA. *Hearing.* Mahwah, NJ: Troll, 1988. The interactions between our sensory organs and the brain are described at a level that is simple enough to help kindergarten/primary-grade children appreciate these marvelous processes.

[*]Starred references, written at the young child's level of understanding, can help teachers who have minimal backgrounds in science to expand their knowledge base.

Cheerful illustrations supplement this excellent information. Paperback.[*] Other books in this question-and-answer series are *Seeing, Smelling, Tasting, Thinking,* and *Touching.*

SUHRL, MANDY. *How I Breathe.* Minneapolis: Carolrhoda Books, 1992. Beginning concepts about how humans and other living things take in and use oxygen are presented simply. Paperback.

SUHRL, MANDY. *I Can Move.* Minneapolis: Carolrhoda Books, 1992. Simple facts about muscles, bones, and joints are presented in a narrative style, humorously illustrated. Paperback.

Stories and Resources About Health Care

ASIMOV, ISAAC, & DIERKS, CARRIE. *Why Do We Need to Brush Our Teeth?* Milwaukee: Gareth Stevens Publishing, 1993. Why and how to brush teeth, and straightforward information about tooth structure and care are the focus of this book for independent readers, and for selective reading aloud to younger children. Also in this series: *Why Do We Need Sleep? How Does a Cut Heal?* and *Why Do Some People Wear Glasses?*

BERGER, MELVIN. *Germs Make Me Sick.* New York: Harper, 1986. Children will be fascinated and relieved to learn about how our bodies fight bacteria and viruses. Paperback. A Reading Rainbow Book.[*]

BERGER, MELVIN. *Ouch!: A Book About Cuts, Scratches, and Scrapes.* New York: Dutton, 1991. Written for older children, it can be read aloud selectively to younger listeners. Well-illustrated to add understanding of the body's healing capacities.

DAVISON, MARTINE. *Robby Visits the Doctor.* New York: Random House, 1992. Part of the series written for the American Medical Association to prepare children for different health-care situations. Written by Davison, the series includes: *Rita Goes to the Hospital, Maggie and the Emergency Room,* and *Kevin and the School Nurse.*

HOBAN, RUSSELL. *Bedtime for Frances.* New York: Harper & Row, 1960. Did Frances and her family get enough rest during a confusing night?

ROGERS, FRED. *Mister Rogers Talks About Going to the Doctor.* New York: Platt & Munk, 1972. The low-key, calmly presented text and photographs prepare young children for visits to the doctor. It may still be on your library's shelves.

UNWIN, MIKE, & WOODWARD, K. *What Makes You Ill?* Tulsa, OK: EDC Publishing, 1993. Capsules of basic information, warmed with relevant illustrations, touch on a wide range of illness topics that interest children deeply. Paperback.

Stories and Resources About Nutrition

DAVISON, MARTINE. *Kevin and the School Nurse.* New York: Random House, 1992. A good story line incorporates a simple discussion of the food guide pyramid. Paperback.

DOOLEY, NORAH. *Everbody Cooks Rice.* Minneapolis: Carolrhoda Books, 1991. A comfortable story told about neighboring families from eight different ethnic groups preparing their special rice dishes for dinner. Delicious recipes are given for this grain-group staple. Detailed illustrations of family life add to this book's usefulness in the integrated curriculum.

ERLBACH, ARLENE. *Peanutbutter.* Minneapolis: Lerner Publications, 1994. Written for older children, the photographs and facts about growing and processing this interesting seed into a favorite food will intrigue pre-readers when selectively read to them. It includes a recipe for making peanut butter from scratch and other smart snacks.

HOBAN, RUSSELL. *Bread and Jam for Frances.* New York: Harper & Row, 1964. Frances goes on an empty-calories eating binge. The problem is resolved humorously and effectively.

KATZEN, MOLLIE. *Pretend Soup and Other Real Recipes.* Berkeley, CA: Tricycle Press, 1994. Nutritious recipes and clear, step-by-step drawings guide young classroom cooks.

MILLER, SUSANNA. *Beans and Peas.* Minneapolis: Carolrhoda Books, 1990. This book provides needed ideas about vegetable proteins as significant protein sources. It includes simple recipes and cross-curricular information about these ancient and universally valued foods.

MORRIS, ANN. *Bread, Bread, Bread.* New York: Lothrop, Lee & Shepard, 1993. Marvelous color photographs by Ken Heyman show breads being made, sold, or eaten in 28 countries.

PATENT, DOROTHY. *Nutrition.* New York: Holiday House, 1993. Basic nutrition is carefully presented and amplified with good photographs and a food guide pyramid. The writing level needs to be paraphrased for younger children.

PILLAR, MARJORIE. *Pizza Man.* New York: Thomas Y. Crowell, 1990. Photographs and a simple text take us from beginning to end of the pizza-making process.

RATTIGAN, JAMA. *Dumpling Soup.* Boston: Little, Brown, 1993. Good foods from many cultures are involved as a Hawaiian family prepares a New Year's feast. Charming watercolors inform and delight. Winner: New Voices, New World Award.

ROBBINS, KEN. *Make Me a Peanut Butter Sandwich, and a Glass of Milk.* New York: Scholastic, 1992. To appreciate what really goes into a simple snack, the author photographed the whole production process starting with harvesting wheat and peanuts, and grazing cows.

SOTO, GARY. *Too Many Tamales.* New York: G. P. Putnam, 1993. A well-plotted story centers on a child's first experience helping make tamales for a family festivity. Grains and meat groups combine in this tamale story.

THOMPSON, PEGGY. *Siggy's Spaghetti Works.* New York: William Morrow, 1993. From wheat field to the dinner table, the whole process for making spaghetti is revealed. Tucking in a bit of pasta history increases the cross-curricular usefulness of this grain-group story.

Poems (Resources in Appendix A)

From *Now We Are Six* by A. A. Milne, read "Sneezles."

From *Egg Thoughts and Other Francis Songs* by Russell Hoban, read, "Egg Thoughts."

From *I Thought I'd Take My Rat to School,* edited by Dorothy Kennedy, read "Lunch."

Fingerplays

Use this fingerplay to encourage handwashing:

This right hand is a very fine hand.
This left hand is its brother
Together they wash with soap and water.
One hand washes the other.
—AUTHOR UNKNOWN

Art Activities

Paper Doll Chain. To recall that people are alike in many ways, fold and cut out paper doll chains. Encourage children to use markers or crayons to make each paper doll different. "Make each one look special, just as each of you is special."

Individuality Collage. As a group project, make a collage poster by using pictures of persons of varying ages and races to emphasize the uniqueness of each person.

Fingerprint Art. Children's fingerprints can create all-over designs and patterns. Use a simple paint stamping pad for prints (see Chapter 4, Plant Life). Suggest using crayons to add stems to print-petaled flowers, strings on fingerprint balloons, and so forth.

Clothespin Dolls. Plain wood clothespins form head, trunk, and legs of dolls; pipecleaners twisted around the "neck" form arms and loop hands. Features can be drawn on the heads with ballpoint pen. Scraps of fabric can be fastened to the body with rubber bands. Compare the flexible and rigid parts of the doll bodies with the bone structure of the human body.

Dramatic Play

Hospital Play. Add paper towel tubes to the props provided for hospital play so that children can listen to heartbeats. Add discarded plastic syringes (sterilized, needles removed) to allow children to play the role of shot-giver.

Grocery Store. Before and after the nutrition experiences, compare the ways that children arrange the food replicas provided for a play store.

Restaurant. Make picture menus from manila folders, using illustrations of wholesome foods marked with large-numeral prices. Provide restaurant workers with pads of paper, pencils, a toy cash register, trays, and food replicas (or pictures cut from cardboard food cartons).

Creative Thinking

Alike and Different Game. Ask one child to look around at the other children to find someone who is like her or him in *one* way. If Elena chooses Jenny because they are both girls, ask Jenny to stand next to Elena and find one way that she and Elena are different. Then Jenny has a turn to find someone who is like her in one way, and so on. End the game with a comment about no two people being just alike in every way. Mention the good thinking and choosing the children did.

A Fantasy Sensory Walk. (Suggested by Hope Jordan.) Tape-record outdoor sounds: twigs breaking, squirrels scolding, rocks splashing as they fall in a puddle, waves breaking, etc. Collect as many objects as possible, such as pine sprigs and cones, rocks, and apple slices. Later tell a story of a walk in the woods to closed-eyed children, who imagine themselves on that walk with you. In a hushed tone question what you are experiencing: "What brushed against my arm? What kind of fruit is growing on that tree?" Pass around the pine sprig, slices of apple, etc., as they fit the story. Ask similar questions while replaying the sounds. Afterward, make a chart listing the senses across the top. On the side, list the events or objects children recall encountering in the fantasy walk (dew on the grass, crow, apple). Ask, "How did we know it was a pine branch . . . a crow . . . an apple?" Let children place a check under the senses that told them what the event or object was, completing the chart.

Creative Movement

Try to find a marionette to demonstrate its stiff movements controlled by strings. Compare the differences between the carved figure and children's bodies. Encourage the children to dance to a record, using their bodies as though someone else controlled their movements with strings. End the dancing by saying that someone has "cut their strings."

Show the children a rag doll, whose body is so flexible that it can be bent and curled in many ways that the marionette and children could not move. Let the children dance like rag dolls without bones or muscle control. Change the movements to those of real people who have muscles to stretch.

Field Trips

Public Health Clinic. If a public facility can be toured, try to have the visit serve the children in a nonthreatening way, such as having children measured and weighed. Be certain that the children are prepared for what to anticipate at the clinic.

A Visit from the Field. Your community emergency medical service squad may be willing to bring a field trip to your school door. Children feel more at ease if a health professional visits them in a familiar setting. A careful look at the emergency vehicle can be fascinating and reassuring. The squad members evoke the awe of heroism, rather than the shadow of anxiety that may be aroused in some children when they meet health-care professionals.

SECONDARY INTEGRATION

Maintaining Concepts

Teachers who are committed to the principle of accepting responsibility for one's own health and nourishment find it easy to use teachable moments to promote this goal. When a respected teacher joins the children in routine hand-washing and toothbrushing, children emulate the patterns readily. When giving first aid to a scraped-knee victim, reassure the child that parts of the flowing blood are already beginning to make repairs so that new skin can grow over the scrape. Encourage children to discuss feelings about medical treatment when they have it and express respect for the way they faced what had to be done to keep themselves healthy.

Teachers have an excellent opportunity to help children appreciate nonsweet foods when they have snacks or lunch with the class. This is a natural time to discuss food and to talk about the food groups represented in the menu. Use the simple rule that everyone (including adults) tries to taste each food being served, allowing (rather than requiring) children to eat as much as they can. Avoid using food as a punishment or as a bribe to force children to eat disliked food. This practice weakens trust between child and teacher and may fail to win compliance from the child as well.

Undue "preaching" of health care and nutrition messages to dictate home practices can become counterproductive. Young children are limited in their ability to change the habits or cultural patterns followed by their families. Therefore, pressing children to conform to healthful standards away from school may result in feelings of guilt or resentment. The common practice of requiring children to report what they have eaten at home each day can threaten feelings of acceptance and reduce self-esteem. Neither practice may result in the goal of translating health-care knowledge into actual health-care practices. Teachers contribute most to this goal when they offer accurate information to children and parents, model sensible eating and health-care habits, encourage children to follow their lead, and support those who try.

Connecting Concepts

1. When children are learning about their sense of vision, think about how light passing through curved glass makes things look larger (see Chapter 15, Light). "This is how glasses help some people see better."
2. When discussing hearing, mention the delicate part inside of each ear that vibrates to let us hear sounds (see Chapter 14, Sound).
3. Identify the calcium that our bones and teeth get from milk and dark green vegetables as the same mineral that makes hard shells for some animals (see Chapter 5, Animal Life) and compresses or combines with other minerals to form certain rocks (see Chapter 10, Rocks and Minerals).
4. When children are learning about the muscles that keep the lungs expanding and contracting, mention that all living things need air to stay alive.
5. Relate muscle development and exercise to simple machines by providing an inexpensive rope and pulley exerciser in the classroom.
6. Recall seed sprouting experiences (see p. 118–120). Each seed contains enough nourishment to start a new plant. That same nourishment also helps all parts of our body grow. There would be enough food to feed everybody in the world if we all would eat more protein from plants and less from meat. (See *Diet For A Small Planet* by Frances Lappe.)
7. Relate the need for clean hands to the need for sanitary food preparation by using this comparison: Seal unrinsed wheatberries in one plastic sandwich bag, together with a damp paper towel. Seal thoroughly rinsed wheatberries in a similar bag. Leave flat, undisturbed for 2 or 3 days. Examine the seeds with a magnifier several days later for clear evidence of mold or bacterial growth.
8. Experiment with two pots of a fast-growing plant such as coleus. Provide only one plant with liquid fertilizer each month. The difference in growth in a few months' time will illustrate how proper nourishment helps living things grow. (Make it clear, however, that nourishing substances for plants can make humans very sick.)
9. For some classes it might be desirable to explore the concepts in this chapter in smaller clusters during the school year. This sequence could be followed: individuality, sensory perceptions, anatomy and physiology, health-care responsibility, and nutrition. The *All About Me* booklet could be developed over this period of time to serve as a source of continuity for the related concepts.

Family Involvement

Two goals of this chapter, valuing one's self and assuming responsibility early in life for one's own health, cannot be achieved without parental involvement. While few parents intend to neglect their children's health, there has been a marked decrease in communicable disease immunization for young children. The number of malnourished children in our affluent country is also discouraging. These problems are being addressed by the American Pediatrics Association campaign to urge prevention of illness through responsible self-care, rather than through reliance upon costly treatment to cure illness. Teachers and families together can encourage responsible attitudes in children. The *All About Me* booklet can spur families' interest in promoting healthy development for their children.

RESOURCES

The Human Body

Allison, Linda. *Blood and Guts: A Working Guide to Your Own Insides.* Boston: Little, Brown, 1976. Accurate information about anatomy and physiology presented in a lively style. Experiments suggested for older children may be adapted for younger children.

Greene, Martin. *A Sigh of Relief: The First Aid Handbook for Childhood Emergencies* (4th ed.). New York: Bantam Books, 1994. Easy, clear information and illustrations of first aid steps and child safety precautions. A logical choice for every school's professional bookshelf.

Johnson, Helen. (1994). The bodyworks: Inside me—another approach to alike and different. *Young Children, 49,* 21–26.

Parker, Steve. *Skeleton.* New York: Alfred A. Knopf, 1988. Outstanding color photographs and art enhance the basic information about the structure and function of the human and animal skeletal systems. London Museum of Natural History Eyewitness Series.

Rockwell, Robert, Williams, R. & Sherwood, E. *Everybody Has a Body.* Mt. Ranier, MD: Gryphon House, 1992. Further explorations and extensions for the classroom.

Stein, Sara. *The Body Book.* New York: Workman Publishing, 1992. Intended for middle school children, this book provides excellent background reading about human anatomy and physiology. It informs without overwhelming the reader.

Van Cleave, Janice. *Biology for Every Kid.* New York: Wiley, 1990. Experiments with lung capacity measurement are included.

Nutrition

Baird, Pat. *The Pyramid Cookbook.* New York: Henry Holt, 1994. This beautiful book takes us through each layer of the food pyramid. It provides sound nutrition background, and lists pyramid equivalents for each recipe. Highly recommended.

Brody, Jane. *Jane Brody's Good Food Book.* New York: W. W. Norton, 1987.

Burns, Marilyn. *Good For You.* Boston: Little, Brown, 1978. Hard facts with a light touch about the digestive system and good nutrition. Interesting experiments for enrichment.

Inglis, Jane. *Fiber.* Minneapolis: Carolrhoda Books, 1993. Good information about an important topic that can be appropriately added to class discussions. Color photographs of old flour mills and high fiber foods.

Inglis, Jane. *Proteins.* Minneapolis: Carolrhoda Books, 1993. Good information and photographs of animal and vegetable protein sources. Includes a nice analogy with house construction to explain the body's need for protein.

Johnson, Sylvia. *Wheat.* Minneapolis: Lerner Publications, 1990. Good photographs and information about growing and harvesting this important member of the grass family.

Lappe, Frances. *Diet for a Small Planet.* New York: Ballantine, 1985. This book clarifies how world food needs can be met by producing more vegetable than meat proteins. Interesting, low-cost recipes are included.

Nottridge, Rhoda. *Fats.* Minneapolis: Carolrhoda Books, 1993. This book presents a balanced view on the appropriate place for fats in human nutrition. A simple experiment to descern the presence of excessive fats in foods, and a calm discussion about maintaining appropriate weight safely are included. Good color photographs.

Walker, Barbara. *Little House Cookbook: Frontier Foods from Laura Ingalls Wilder's Classic Stories.* New York: Harper Collins, 1979. New York Times Outstanding Book.

Additional Resources:

Delicious Decisions. Food Choices to Grow On. A complete curriculum to teach the Food Guide Pyramid for grades K–6. Separate posters, a blackline master of a three-dimensional pyramid, and packets of stickers can be ordered from a free catalog of nutrition education materials. Write to the Dairy Council of Wisconsin, 999 Oakmont Plaza Drive, Suite 510, Westmont, Illinois 60559.

Mail Order Grains Sources:

Arrowhead Mills
110 S. Lawton
Hereford, TX 79045

Meadowlark Grain Company
P.O. Box 93
Hingham, MT 59528
(Wheat only)

What Is Air?

Children become acquainted with hundreds of substances through sensory encounters. Air intrigues children because it is an invisible, ever-present substance that is not directly encountered by the senses until it affects something tangible. The activities in this chapter are designed to reveal the following concepts:

- Air is almost everywhere.
- Air is real; it takes up space.
- Air presses on everything on all sides.
- Moving air pushes things.
- Air slows moving objects.

In the experiences that follow, children will find air in empty containers and become aware of the air that they breathe. They will feel air enclosed in something, see how it occupies space, and note how air pushes on everything. Also, they will enjoy making things for moving air to push.

Introduce the topic of air to children by holding up your hands, cupped together, suggesting, "There is something important inside my hands. You may peek in, but you won't be able to see what it is. The important stuff is *invisible!* We'll be finding out more about it."

CONCEPT: Air is almost everywhere.

1. What comes out of an empty can?

LEARNING OBJECTIVE: To develop a beginning awareness that air is present almost everywhere, even if unseen.

MATERIALS:

Empty metal can, #303

Nail, hammer

Dishpan

SMALL-GROUP ACTIVITY:

1. Ask children to examine the can. Do they notice anything in it?
2. Use a hammer and nail to punch a hole in the bottom. Let a child do it, if possible.

131

Water

Plastic aprons for participants

Splash catcher (Inflatable wading pool is fine. Use on floor, inflated, to confine spills. Put pan of water in center of the pool. Children stay outside the pool.)

3. Ask: "If there is something in the can, could we push it out through this hole? Let's see."
4. Invert can; slowly push it into the water. Invite a child to hold his hand just above the nail hole. What does he feel being pushed out of the hole? A stream of air? "Could something real be in that empty can . . . something real that we can't see?"
5. Let the children experiment.

2. Can we really empty an open container?

LEARNING OBJECTIVE: To extend awareness that air is present almost everywhere, even if unseen.

MATERIALS:

Empty clear plastic shampoo bottles or tubes, uncapped
Downy feathers, milkweed fluff, or bits of tissue paper

SMALL-GROUP ACTIVITY:

1. Ask children to examine the bottles or tubes. Do they notice anything inside?
2. "Point the bottle top toward your chin and squeeze the bottle. Do you feel something? What? See if you can squeeze it out. Keep trying."
3. Offer a feather to each child. "If your bottle is really empty, nothing will come out to push the feather from your hand. See what happens if you point the bottle toward the feather." (This activity can be disorderly unless you set limits for a feather-blowing space.)

3. Is there air inside us?

LEARNING OBJECTIVE: To become aware of the air we breathe.

Group Activity: Suggest that all the children take a deep breath, shut their lips tightly, and pinch their nostrils shut for as long as they can. "What happened? Why did you let go of your nose? What did your body seem to need so much that you had to let go? Let's try again."

"Our bodies need the part of air called oxygen. We need to breathe air into our bodies about every 6 seconds because it is used up so quickly inside of us." Guide the discussion to bring out the idea that all living things must have air to stay alive. Discuss the danger of climbing into boxes with heavy lids or other enclosed spaces that cannot be opened from the inside. Give examples.

Give each child an unfolded tissue to hold near his or her nose and mouth. "Let's see what happens to the tissue when you bring air into your body and then when you push this air out. Try it with your lips closed; now with your lips opened a bit. Are there two ways for the air to come in and out? Let's see if we can tell where the air is going inside our bodies." Ask children to take turns lying on the floor or putting their heads on the table so others can watch their chests go up and down as they breathe.

Read *Whistle for Willie* by Ezra Jack Keats. Comment about how we breathe air and box-play safety.

CONCEPT: Air is real; it takes up space.

1. Can we feel the air inside a bag?

Brook can hold a cork under water without touching it.

LEARNING OBJECTIVE: To develop a beginning awareness that air has substance.

MATERIALS:

Plain small plastic bags and covered wire closures (or small zip-top bag)*

Drinking straw

Transparent tape

Covered wire closures

SMALL-GROUP ACTIVITY

1. Ask: "Is the table real? How can you tell? If your eyes were closed could you still tell that the table is real? Is feeling something a good way to know that it is real?"
2. Let children blow into the prepared bags. "What are you blowing into the bag? Look in the bag. Feel it. Do you think something real is in there?"

*Balloons may be used instead of the sandwich bag and straws. However, the more transparent the container of air, the more apparent the concept that invisible air is taking up space inside. We can feel air inside a closed container, we can see the container enlarge, but we cannot see the air.

GETTING READY:

Prepare a blowing bag for each child participating by gathering the top of a bag to form a neck, inserting a straw in the neck, and securing the neck of the bag to the straw with tape.

3. Remove straw and tie with wire closure, or zip seal so to children can keep their bags full of air to feel.

Children should leave the plastic bags at school so that younger children at home will not pick up the idea of playing with plastic bags.

2. Can a glassful of air push water away?

LEARNING OBJECTIVE: To extend the awareness that air has substance.

MATERIALS:

Clear plastic bowl (corn popper dome is fine) or 5" (12 cm) deep glass mixer bowl, if it can be safely used

Pitcher of water

Cork

Clear plastic tumbler

Splash catcher, aprons

GETTING READY:

Inflate wading pool splash catcher. Put bowl in the center.

SMALL-GROUP ACTIVITY:

1. Put cork in the bowl. "The cork is touching the bowl now. What will happen if we pour water into the bowl? Let's see."
2. Fill half the bowl with water. "Notice if the cork is touching the bottom of the bowl now."
3. Ask: "Can someone put the cork back on the bottom?" Let children try to do this. "Will it stay if you let go?"
4. "See if something in the tumbler will push water away so that the cork can stay on the bottom of the bowl." Invert tumbler and push straight down to the bottom. "What could be inside the glass pushing the water away? Do you see anything?"
5. Let the children experiment.

FIGURE 7–1

Group Discussion: Ask the children to catch some air in their cupped hands. "Poke your nose into your hands. Can you smell the air? Peek in. Can you see it? Can you hear it? Try to taste it." (Children love to hang their tongues out for research purposes.) Ask those who have experimented how they know that air is real. Children are proud of their new information and are willing to repeat a discussion like this one. Next time, let a hand puppet be the skeptic who doesn't believe that invisible things can be real. The children will be happy to convince him otherwise.

3. Can air in a bag push something over?

LEARNING OBJECTIVE: To extend the awareness that air has substance.

MATERIALS:

Large, sturdy paper bag

Wire twist closure, or tape

Bicycle tire pump, or bellows-
type air mattress pump

Heavy book or block

GETTING READY:

Gather bag top into a neck
and fasten with a wire closure or tape.

Put bag on floor and stand
book upright on it. Put the
pump near the bag (Figure
7–1).

SMALL-GROUP ACTIVITY:

1. Demonstrate the pump if children don't know
what it is or how it works. Look at the air intake
hole where air comes in to be pushed out by the
pump. (Remove intake valve cap from bellows
pump.) Feel air being pushed out.
2. Ask: "Do you think a bag full of air could push that
book over? How could we find out?"
3. Insert the end of the pump hose into the bag.
Tighten the closure, or simply hold bag firmly to
hose while children take turns pumping air into it,
pushing the book over. "Could air be doing that?
Do you think air is real stuff, even though we can't
see it?"

Note: Children love to demonstrate how strong they are. Using the pump is a good medium for this self-concept enhancement.

CONCEPT: Air presses on everything on all sides.

1. Can air keep water in a tube or straw?

LEARNING OBJECTIVE: To notice the effects of air pressure on liquids in a small tube.

MATERIALS:

Baster (kitchen tool)

Medicine droppers (plastic
or thick glass tubes are
safest)

Small containers of water
for each child

Pint jar half full of water

Sponge, pan, and news-
papers or splash catcher
for cleanup

SMALL-GROUP ACTIVITY:

1. With children, examine baster. Is the tube empty?
Squeeze bulb to feel what comes out.
2. Put baster in jar of water. "What do you see when I
push air out of the baster? Watch the tube when I slowly
let go of the rubber bulb on top. What is happening?"
3. Hold up the filled baster. "What's happening to the water now? What could be keeping the water in? Perhaps
something invisible is in the baster with the water."
4. Squeeze bulb to push out water. Release bulb to pull
in air. "What's happening now? Try it."

Julie squeezes the air: "Now it's coming out!"

Drinking straws, cut in half

5. "Now let's try the same thing with straws." Insert a straw in water. Put your index fingers over the top and bottom of the straw and lift it from the water. Keep finger on top and remove finger from bottom. Most of the water will stay in the straw. Why? (Air pressure keeps it in.) See Figure 7–2.
6. "Nothing can get into the straw from the top with my finger on it. Watch as I let some air push into the straw." Remove finger from top. Water is pushed out of the straw. Let children experiment.

Note: Some children may believe that a dropper is filled with water if drops just cling to the outside. Help them hold the dropper tube in the water *while* releasing the bulb.

CONCEPT: Moving air pushes things.

Introduction: Ask the children to wave their hands back and forth in front of their faces. "Can you feel something on your skin? When air is moving you can feel it. You don't notice it when it isn't moving." Explain that we feel air moving against us in the same way whether we are moving when the air is still, or whether air is moving by itself (as wind) when we are still.

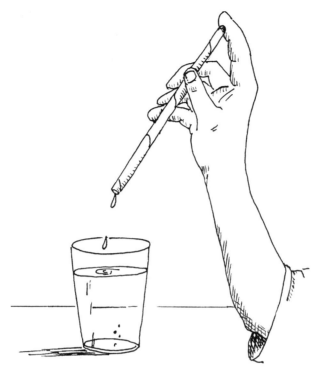

FIGURE 7–2

1. Does moving air push on us?

LEARNING OBJECTIVE: To experience the pushing effects of moving air.

MATERIALS:

Sheets of newspaper

Used adding machine tape

Pinwheels or kites

OUTDOOR ACTIVITY:

Give children sheets of paper outside. "As you stand still, will paper stick to the front of you if you don't hold it there? Now see what happens to it when you run as fast as you can. Do you have to hold the paper to press it to yourself? What could be holding it?" Let children enjoy running with paper tape streaming behind them. Let them try pinwheels while standing still and running. Try the kite on a calm day and on a windy day. (Launch the kite as you move toward the wind.)

Read about a struggle to fly a kite in *Days with Frog and Toad* by Arnold Lobel.

2. Can we make a glider drift on moving air?

LEARNING OBJECTIVE: To notice that moving air can carry a glider launched by a child.

MATERIALS:

A sheet of typing paper for each child

GETTING READY:

Draw a lengthwise line down the center of the paper.

Draw parallel lines 1 1/2" (4 cm) from the center line.

Draw a line 3/4" (2 cm) from the bottom across the paper.

SMALL-GROUP ACTIVITY:

1. Show the children how to fold up 3/4" (2 cm) from the bottom of the paper. Continue to fold this amount four times.
2. With the cuff on the outside, crease the paper and the folded cuff lengthwise on the center line. Now fold back each side on the lines 1 1/2" (4 cm) from the center fold, making wings. (See Figure 7–3.)
3. Launch glider with cuff in front, holding the center. If the glider always nosedives, reduce the weight of the front end by unfolding one fold of the cuff. Push the glider ahead by keeping the hand and arm *straight,* not sweeping downward, when launching the glider.

Note: Primary-grade children may want to make more elaborate gliders. See *Fabulous Paper Airplanes* by Richard Churchill.

CONCEPT: Air slows moving objects.

Group Discussion: Show the children two pieces of typing paper: one piece lying flat and the other piece tightly crunched into a ball. Ask: "If I hold these two things up as high as I can reach and then let go of them, what will happen to them?" (Wait for responses.) "Do you think that both things will fall in just the same way? Let's find out." Repeat the action as needed to allow the children to carefully observe and report. "Which one fell more slowly? Can you think of anything beneath the wide paper that might have pushed against it and slowed its fall?" Mention that everything that moves above ground or above water must push air aside as it moves. Big things push more air aside than do small things.

1. What does a parachute do?

LEARNING OBJECTIVE: To compare the ways air slows the fall of a weight with and without a parachute.

MATERIALS:

For each child:

1 wooden thread spool or small toy figure[*]

12" (30 cm) square of closely woven, light fabric

4 feet (1.2 m) light string

Extra wooden spool or toy figure

Large needle

SMALL-GROUP ACTIVITY:

1. Allow children to help make parachutes if they can tie. Children can crayon designs on fabric before strings are tied.
2. If a safe stairway is nearby, let one child go up and reach through banisters to drop the extra spool and the parachute. Compare results.
3. Fold parachute fabric and wind with the strings on spool. Let children throw them as high as they can into the air outdoors. Add weight to the spool if parachutes fail to open.

[*]The spool must be heavier than the fabric. Tie on metal washers or nuts if more weight is needed.

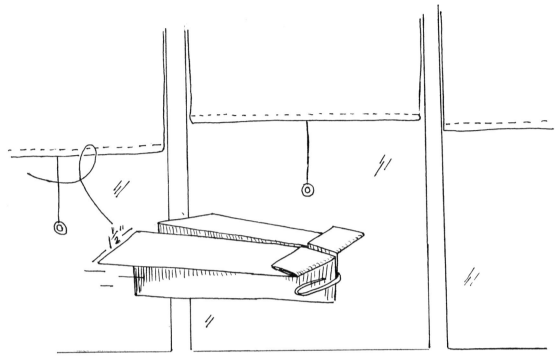

FIGURE 7–3

GETTING READY:

Sew a 12″ (30 cm) piece of
  string to each corner of fabric.

Hold parachute by strings,
  fabric corners even, as
  shown in Figure 7–4. Knot
  strings together 7″ (18 cm)
  from corners.

Tie spool to string ends.

PRIMARY INTEGRATING ACTIVITIES

Math Experiences

1. Practice relaxation breathing to focus attention on the lungs holding air. Say, "Breathe in, one, two; breathe out, one, two." Repeat several times. This helps children feel the air in their bodies and also, done rhythmically and slowly, it helps children relax.

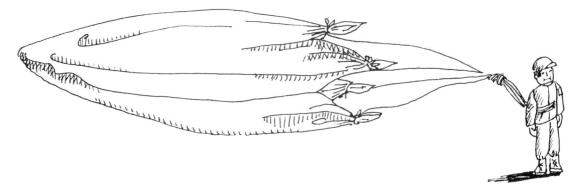

FIGURE 7–4

2. Ask children what happens to the air inside a balloon over time. Tape an inflated balloon to paper on the bulletin board. Observe what happens to it over several days. Record its shrinkage by tracing a circle around the balloon. Ask children to speculate on the event. Refer to the concept that air presses on everything.

Music (Resources in Appendix A)

After children have finished clapping the rhythm of a song they enjoy singing, compare two styles of clapping. Listen to the sound made by clapping with palms open and flat, then to the sound made by clapping with slightly cupped hands. What is caught between the cupped hands to change the sound?

Sing this song to the refrain of "Goober Peas" (in *Folksong Festival*, Appendix A):

> *Air, air, air, air*
> *air is everywhere.*
> *We can't taste or see it,*
> *but we know it's there.*
> *—J. H.*

Use appropriate tasting and peering gestures to add fun.

Stories and Resources for Children

ARDLEY, NEIL. *The Science Book of Air.* San Diego: Harcourt Brace Jovanovich, 1991. Interesting experiments and clear explanations are illustrated with vibrant color photographs.

BRANDT, KEITH. *Air.* Mahwah, NJ: Troll, 1985. Intended for elementary-grade readers, portions of this book can be read aloud to pre-primary and primary-grade children. Paperback.[*]

BRANLEY, FRANKLYN. *Air Is All Around You.* New York: Harper & Row, 1986. Simple air concepts and experiments, charming illustrations. Paperback.

BURLEIGH, ROBERT. *Flight.* New York: Philomel, 1991. Older children will appreciate this story of Lindberg's courageous first transatlantic flight. Illustrations put the reader in the cockpit.

CALHOUN, MARY. *Hot Air Henry.* New York: Morrow, 1981. Henry the cat has unexpected adventures in a hot-air balloon. Exciting illustrations by Erick Ingraham clearly show how the balloon's movement is controlled. Paperback.

CHURCHILL, E. RICHARD. *Fabulous Paper Airplanes.* New York: Sterling, 1992. This book of-fers directions for making 29 paper airplanes, many easy enough for primary-grade children to manage. Paperback.

GLOVER, DAVID. *Flying and Floating.* New York: Kingfisher Books, 1993. Simple directions are given for making a paper hot-air balloon and pinwheel, an easy kite, and a helicopter rotor. Paperback.

GRIFFIN, MARGARET & RUTH. *It's A Gas!* Toronto: Kids Can Press, 1993. Experiments and information add to older children's knowledge about air, including the mixture of gases that make up air, and how our bodies use only the oxygen from the air we breathe. Paperback.[*]

KEATS, EZRA JACK. *Whistle for Willie.* New York: Viking Press, 1963. Peter hides in a box that he can get out of easily. He also learns how to blow air out of his lips to whistle.

LOBEL, ARNOLD. *Days with Frog and Toad.* New York: Harper, 1990. Toad and Frog persist, in spite of discouragement, and finally are able to fly their kite in the wind. Paperback.

MICHAEL, DAVID. *Making Kites.* New York: Kingfisher Books, 1994. Careful instructions are given for making and flight-testing a variety of interesting kites, including one made from a plastic raincoat! A good book to recommend to families seeking a project to share with children. Paperback.

MILNE, A. A. *Winnie-the-Pooh.* New York: Dutton, 1961 (Chapter 1). Read this after children have made parachutes. Ask whether a small balloon could really slow Pooh's fall.

REY, H. A. *Curious George Gets a Medal.* New York: Scholastic Book Services, 1957. After a space flight, George drifts to earth under a parachute.

SUHR, MANDY. *How I Breathe.* Minneapolis: Carolrhoda Books, 1991. Simple information for young readers on the composition of air, and how our bodies need and use oxygen.

TAYLOR, BARBARA. *Up, Up and Away! The Science of Flight.* New York: Random House, 1992. Clear information and directions for simple activities cover several properties of air, as well as flight principles. Paperback.

WADE, ALAN. *I'm Flying!* New York: Alfred Knopf, 1990. A boy fantasizes that a box of weather balloons and a tank of helium gas let him float away from earth's gravity pull to explore the world. A Reading Rainbow book. Paperback.

WHITE, LAURENCE. *Science Toys and Tricks.* New York: Harper & Row, 1990. The impressive upside-down glass of water experiment demonstrating air pressure, as well as a simple aerodynamics principle, are in this collection. Paperback.

WYLER, ROSE. *Science Fun with Toy Boats and Planes.* New York: Simon & Schuster, 1986. Includes good information about the lifting power of air streams and paper glider construction.

Poems (Resources in Appendix A)

The wind is a favorite topic in poems for children. The following are recommended:

> In Poems Children Will Sit Still For by de Regniers:
> "Who Has Seen the Wind, " by Christina Rossetti
> "Wind Song," by Lillian Moore
> In *Now We are Six* by A. A. Milne:
> "Wind On the Hill"
> In *Poems to Grow On* by Jean Thompson:
> "The Kite," by Harry Behn

---

[*]Starred references, written at the young child's level of understanding, can help teachers who have minimal backgrounds in science to expand their knowledge base.

Fingerplays

Seeds with silky wings are scattered by moving air (see Chapter 4, Plant Life). Recall the milkweed blowing that children may have enjoyed in plant life experiences, or try to find a milkweed pod to open outdoors now if it is a new idea.

Baby Seeds

In a milkweed cradle	(cup hands)
Snug and warm	
Baby seeds are hiding	(make yourself small)
Safe from harm.	
Open wide the cradle	(open hands)
Hold it high.	
Come Mr. Wind	(blow in the hands)
Help them to fly.	

—AUTHOR UNKNOWN

Art Activities

Easel Painting. Cut newsprint into the shapes of things that move on the air, such as birds, balloons (tape a dangling string to the back), butterflies, and kites. Let each child choose a shape to paint with a design of his or her own creation.

Collage. Use airborne nature materials such as feathers (to represent birds); milkweed fluff and seeds; and maple tree seeds. Cut some actual air bubbles from sheets of plastic packing material to glue on the collage.

Paper Fans. Make paper fans to move air. They may be accordion-pleated sheets of paper that the children decorate, or they may be small paper plates stapled to Popsicle sticks.

Gliders. Make simple gliders and decorate with crayon designs.

Air Pressure Art. With medicine droppers, make use of air pressure as an art medium. Let children use the droppers to make designs by dropping diluted food coloring onto absorbent paper. A fleeting, fascinating art form—snow painting—also makes use of droppers and food coloring. Gather a big bucket of snow. Store it on the windowsill. Quickly pack individual plastic meat trays with snow, ready to receive the drops of color. Have another bucket ready for the sloshy end results.

Blow Painting. Do blow painting with plastic straws. Children can use straws and air pressure to pick up tempera paint from jars and drop it on smooth-surfaced paper. The puddles of paint can then be changed in shape by blowing on them through the straws.

Dramatic Play

Service Station. Contrive a gasoline pump from an oblong carton and a length of garden hose. Help the children create a car with chairs, hollow blocks,

Jamal experiments with air-powered painting.

and a paper-plate steering wheel (or a real one). Add an old inner tube and a bicycle tire pump for the repairworkers to inflate. A ring of keys and pliers add fun.

Boat. Help the children create a block and plank boat, with two planks angled together to form a prow. Let them blow up a plastic ring or pump up the inner tube for a life preserver.

Plane. Angled planks form the nose of the plane, hollow blocks form the tail and sides, two more planks stretch out to form the wings. Children who have flown can contribute ideas about oxygen and air to the play.

Housekeeping Play. Try to borrow a child-size inflatable chair to add interest to this play. Let the children beat up a detergent suds "souffle" at the sink by using a wire whisk or eggbeaters. Comment on the amount of air their strong muscles are beating into the dish.

Other Air-Related Play

1. Blow soap bubbles, using straws in cups of water and detergent. For younger children use baby shampoo, which won't irritate eyes, instead of detergent.
2. Let children take turns pumping up a sturdy air mattress to use as a tumbling mat for somersaults.
3. Pump up an inner tube and put it across the seats of two facing chairs to use as a target for foam ball tossing.

4. Let moving air make paper dance. Cut a spiral from a sheet of paper to make one long, curving strip. Dangle it from the ceiling by attaching it to a long thread. Let children fan the air with a card to make the spiral twirl. Create a fantasy effect by dangling several spirals. Circulating air will keep them gently moving.

Creative Movement

1. Let children pretend to be balloons that are blown into large shapes. Suggest that they make themselves into flat, limp shapes on the floor. "Now I'm blowing, and blowing, and blowing some more. And you are getting bigger and bigger." When the shapes have been stretched as large as they can be, suggest that you will pretend to prick each imaginary balloon shape. Each shape collapses into a limp piece of rubber on the floor as the air goes out of it.
2. Children can move like birds or butterflies soaring on moving air. They might enjoy holding a feather in each hand to become a bird with outstretched-arm "wings."
3. Some good ideas for leading creative movement about things that float on air can be found in *Creative Movement for the Developing Child* by Clare Cherry (see Appendix A).

Creative Thinking

Classifying Game. Find pictures of things that move with the help of air (sailboats, planes, kites, balloons, birds, bats, butterflies, and so forth) and pictures of things that move without the help of air (trains, steamships, motor boats, caterpillar tractors, roller skates, and so on). Mount the pictures on cardboard and put them in a box with *yes* or *no* marked inside the lid and the bottom section, respectively. Use the pictures as a game with the whole class or as a quiet table game for a few children. (Pictures of cars and other things that use inflated tires could be classified either way. Air does not push a car along, but a car with a flat tire certainly won't go very far. Let the children decide the issue after discussing it.)

After singing the "Air" song, let the children enjoy thinking of funny places where air creeps in. Start off the game with places like: inside your shoe, your ear, the holes in Swiss cheese, beneath a mosquito's wings . . .

Food Experiences

If an electric mixer is available, make whipped milk topping. Explain that the beater is stirring many tiny air bubbles into the mixture. Watch it change and expand.

> Combine: 1/2 cup instant dry skim milk
> 1/2 cup *cold* water

Beat at highest speed for 4 minutes.

> Add: 2 tbsp. sugar
> 1 tsp. vanilla

Beat at highest speed until mixture stands in peaks, then about 4 minutes longer. Serve a mound of whipped milk to each child. Edible science is impressive. Children really remember that air was part of the tidbit they enjoyed eating.

Field Trips

Airport Visit. If this is possible, keep the science focus simple. There will be too many exciting things to see to make lengthy explanations of flight principles bearable for the children. Be sure that any adult who offers a guided tour of the facility understands this.

Service Station Visit. Make arrangements with the station manager beforehand to have a demonstration of the tire changer and to allow children to feel air from the tire pump. This can be done in connection with a study of simple machines (Chapter 13) if the station is too far away to warrant two trips.

SECONDARY INTEGRATION

Maintaining Concepts

1. Make casual comments or inquiries about the function or presence of air when suitable opportunities arise. When outdoors with a class, observe planes or birds flying overhead, or leaves, seeds, or litter blowing past.

2. Suggest that children fill their mouths with a giant air bubble when the class needs to walk quietly through school corridors. Puff up your own cheeks to model this funny way to stay quiet. It's more positive than saying, "Be quiet!" Air takes up talking space.

3. Use medicine droppers to water tiny seedlings, give moisture to insects in jars, and for art projects. If necessary, remind children to push the air out of the dropper while it is still in the liquid. Can they see the air bubbling into the liquid?

4. If you need to open a vacuum-sealed container, such as using a punch-type opener with a can of juice, let the children listen closely for the hiss of air rushing into the can. When using the bellows-type step pump, show the rubber cap on the air intake valve. Push the air out so the bellows are flat. Remove the cap so that air rushes in with a hiss and watch the bellows expand dramatically.

Connecting Concepts

1. Relate water concepts to air when investigating evaporation. Water droplets, or *vapor,* become part of the air. Speak of fog and clouds as very wet air.

2. Relate plant propagation (seed scattering) to moving air.

3. Relate sound to air concepts. Sound travels through air; moving air causes some things to vibrate and make sounds; air vibrates when enclosed in a column like a flute or a whistle (see Chapter 14).

4. Point out the crucial role of air in ecological relationships. Air is one of the non-living substances that all living things depend on to stay alive. Air that is spoiled by pollution makes problems for all living things. *There is no way to make new air.* It is used over and over again, so people must find ways to keep air clean.

Family Involvement

A newsletter to parents can suggest ways to support and add to the air experiences at school. When families are out together, they can watch for soaring birds, planes, and clouds being pushed along by moving air. They can point out how air is used in their homes to keep people comfortable: heated in winter, passed through coolers in summer. Notify families if there are exhibits in your area where children can see model or real gliders, small planes, windmills, or weather vanes. Suggest showing any dried-out baked goods or shriveled vegetables that have been left out in the air too long. Talk about air picking up moisture from everything.

RESOURCES

HANN, JUDITH. *How Science Works* (pp. 115–123). Pleasantville, NY: Reader's Digest, 1991. Basic information about the properties of air and air pressure and experiments.

LEVENSON, ELAINE. *Teaching Children About Physical Science* (pp. 71–73). New York: Tab Books, 1994.

MARKLE, SANDRA, *Exploring Spring*. New York: Avon, 1990. To understand how kites fly and for the kite-flying safety tips, first-time kite flyers should read "Flights of Fancy" before heading outdoors with the class.

What Is Water?

All living things require water to survive. Water is also splendid for pouring, splashing, and spraying. Most children enjoy learning about this vital part of our environment. The experiences in this chapter will explore the following concepts:

- Water has weight.
- Water's weight helps things float.
- Water goes into some things, but not others.
- Some things disappear in water; some do not.
- Water goes into the air.
- Water can change forms reversibly.
- The surface of still water pulls together.

It is hard to imagine sparkling, transparent water having weight. Once children recognize that water does have substance, they understand more readily how water can hold things up.

In the following experiments, children experience the weight of water, note that water is heavier than air, and explore buoyancy. Other experiences deal with water absorption and repellency, objects that dissolve in water, evaporation, condensation, freezing, melting, and surface tension of water.

CONCEPT: **Water has weight.**

---

1. Does water feel heavy?

LEARNING OBJECTIVE: To experience firsthand that water has weight.

MATERIALS:

Outdoors when it's warm:

Small wading pool

Large bucket

SMALL-GROUP ACTIVITY:

1. Pass pitcher of water for children to dabble fingers in. Ask: "Do you think this clear-looking water could be heavy? Let's find out."

Large pitcher or milk cartons

Water

For indoor use add:

Rubber boots

Plastic smock

GETTING READY:

If outdoors, have children remove shoes and socks and
  roll up trouser legs.

If indoors, have children
  wear boots and smock.

2. Let children take turns standing in the pool and
   holding the empty bucket.
3. Let other children pour water into the bucket until
   it becomes too heavy to hold. Conclusions about
   the weight of water are easily formed and communicated by this direct investigation.

2. Do things weigh the same when dry and when damp?

LEARNING OBJECTIVE: To observe that the weight of a material increases when water is added to it.

MATERIALS:

Part 1

Inexpensive postage or calo-
  rie-counting scale

Matching set of small cups
  such as film canisters, spray
  can tops, etc.

Small pitchers

Water

Sponges, newspapers

Part 2

Tiny scoops

Sand

Pan for wet sand

GETTING READY:

Spread newspaper on a table.

Cover scale markings with
  masking tape, so the indica-
  tor points only to ounce
  numerals.

Make a simple chart for children to keep weight records
  of dry and wet material.

SMALL-GROUP ACTIVITY:

1. Put an empty cup on the scale. Ask: "What does a
   scale do? Let's see what happens if we pour water
   into the cup. What do you think it means when the
   pointer moves down? Which is heavier, a cup of air
   or a cup of water?"
2. Put dry sand in two matching cups. Weigh each to
   be sure they are equal.
3. "Try pouring a bit of water in the sand cup on the
   scale. Where is the pointer now? Why do you think
   it moved?" Compare weights of the wet and the
   dry sand. "Do they weigh the same now?"
4. Let children measure, weigh, and add water. Some
   may wish to try each set of cups to verify results.
5. Let children record dry and wet sand weight on the
   chart.
6. Read the chart results to the whole class after the
   experiment. What was added to the damp sand
   that made it heavier?

CONCEPT: Water's weight helps things float.

1. Will some things float while others sink?

LEARNING OBJECTIVE: To compare which objects, among a collection varying in dimension and weight, will be supported by water and which will not.

MATERIALS:

Dishpans of water

Small wading pool

Plastic aprons

Assorted objects: rocks, sticks, bath toys, corks, washers. Try to find like objects of different materials: ping-pong ball/ golf ball; toy plastic key/ metal key

Old kitchen scale, postage scale, or balance

Two trays

GETTING READY:

Place pans of water in the wading pool.

Tape card marked *float* on one tray and *sink* on the other.

SMALL-GROUP ACTIVITY:

1. Let children enjoy relaxed play with the materials.
2. Ask: "Does the water hold some things up on top? All things? Why do you think this ball stays on top and that ball sinks down?"
3. Encourage children to weigh objects on the scale to confirm that similar objects can differ in weight and thus can act differently in water.
4. Put out *sink* and *float* trays. Suggest classifying objects and placing them in correct trays.

2. Can air help some things float?

LEARNING OBJECTIVE: To discover whether air inside an object helps the object float on water.

MATERIALS:

Dishpans of water

Small wading pool

Plastic aprons

Foot-long (30 cm) pieces of clear garden hose

2 corns to fit each hose

Many small plastic bottles (prescription vials *with* caps)

GETTING READY:

Fit corks securely into hose ends.

Place pans of water in the dry wading pool.

SMALL-GROUP ACTIVITY:

1. Remove one cork from a hose. Submerge it in water. Let it fill almost to the top with water. Replace cork. "What is inside the hose now? Is air or water on top? Which do you think will be on top if the hose is turned upside down? On its side? Try to find out."
2. Drop two capped bottles on the water. "What might happen if you take the cap off one bottle? Do you think it will still float? Find out. What could be in the bottle that still has a cap on it?"

Read *The Story About Ping*, by Marjorie Flack, in which the houseboat boy is kept afloat in the Yangtze River by a barrel of air fastened to his waist.

Phillip feels and sees that cloth sprayed with water won't keep his arm dry.

CONCEPT: Water goes into some things, but not others.

1. What things will absorb water?

LEARNING OBJECTIVE: To observe which materials among a collection will absorb water and which will repel water.

MATERIALS:

Part 1

Plastic aprons

Small dropper-top medicine bottle for each child

Tiny funnel and baster

Plastic meat trays

Part 2

Test materials: sponges, wood, cotton, paper towels, tissues, dried clay, stones

SMALL-GROUP ACTIVITY:

Part 1

1. "What happens when you put drops of water on different things in your tray?" Listen to the children's ideas. Define the term *absorb*.
2. Give a small cube of dry sponge to each child. Put in a dry place. "Watch closely; something will change fast." Squeeze water from baster onto the sponge.

Part 2

1. Move from child to child with a piece of plastic, the dry fabric, and the spray bottle.

6" (15 cm) squares of old cotton fabric

Plastic sheeting or rubberized fabric pieces

Waxed paper, scraps of leather

Smooth feathers

Spray bottle

Water

GETTING READY:

Let children fill bottles with water.

Arrange a tray of test materials for each child

2. Drape fabric over child's arm. Spray with water. "What do you feel?" Repeat with plastic. "Which material would you want for a raincoat? Why?" (This firsthand experience with water repellency intrigues children.)

3. Give each child a feather to hold while you carefully place *one* drop of water on the feather. Listen to the responses. Talk about how birds stay dry in the rain. Let children experiment with water drops, feathers, and leather scraps.

From now on, use the term *absorb* whenever a spill must be wiped up. Let the children use the sponge to absorb spills. Children feel more in control of the situation if they can undo what they have done accidentally.

CONCEPT: Some things disappear in water; some do not.

1. Which things dissolve in water?

LEARNING OBJECTIVE: To observe which materials among a collection will dissolve in water and which will not.

MATERIALS:

Muffin pans and plastic ice cube trays with separated molds, or plastic prescription vials

Pitcher of water

Assorted dry materials: salt, sand, cornstarch, flour, fine gravel, seeds, cornmeal

Spoons for dry materials

Salad oil

Small screw-top bottle

Plastic aprons

Newspaper

Sponge

Sticks for stirring

Cleanup bucket

SMALL-GROUP ACTIVITY:

1. "See what happens when you put a little salt in one of your pans of water. Stir it. Can you see it? Feel it? Where is it?"

2. "Dip a finger in the pan. How does the water taste? The salt is still there, but it is in such tiny bits now it can't be seen. It *dissolved* in the water."

3. "Try the other materials; put each in its own pan of water. Find out which ones dissolve."

4. Half fill the bottle with water. Add some oil. Cap securely. Let the children shake it. "Does it seem to dissolve? Let it stand a while. Where is the oil now?"

GETTING READY:

Spread newspapers on work
  table.

Half fill pans with water.

Group Experience: Mix some sand, dirt, and gravel with water in a pint jar. Let it stand undisturbed for a day or more. Check the jar. Is the water still muddy looking? Did the sand and dirt really dissolve? Which is on the bottom?

CONCEPT: Water goes into the air.

1. Will warm air and moving air take up water?

LEARNING OBJECTIVE: To observe the results of air taking up water and to participate in hastening
                    the process.

MATERIALS:

Two trays

Pans of water

Paper towels

Blackboard or other dark
  smooth surface

Cardboard or paper fans

Hand-held hair dryer

Paper

GETTING READY:

Check safety of the dryer
  cord position

Let children make folded paper fans.

If possible, use the hair dryer
  close to floor level to minimize tripping on the cord or
  dropping the dryer.

SMALL-GROUP ACTIVITY:

1. "If we wet some towels, squeeze them, and
   spread them out on these trays, what do you
   think will happen to them? Let's see." Put one
   tray in a sunny or warm place, the other in a cool,
   darker place. Check them in half an hour.
2. Meanwhile, show children the hair dryer; let them
   feel warm air coming from it. "What do you feel?
   This warm air can help us understand what is
   happening to the towels."
3. Let children dip a finger in water, then trace their
   names on the blackboard. "Blow warm air on your
   wet name. Watch closely to see what happens to
   the water. Where did it go? Only air touched it."
   The water has gone into the air in such tiny drops
   that it can't be seen anymore. When air takes up
   water, we say the water *evaporates.*
4. "Write your name with water on the blackboard
   again. Fold a paper fan and wave it near your
   name to make a breeze. Where does the water go?
   Let's see what happened to the paper towels.
   What happened to the wetness?"

Group Experience: Early in the morning, put measured amounts of water in three pie pans.
Place pans where children can see but not touch them. Have one in the sunlight, one in the
shade, and one close to a heat source. Check several times during the day to note changes.
If the weather is humid, you may need to continue to check the next day. What happened?

Mona tries her own idea for making water change form faster.

CONCEPT: **Water can change forms reversibly.**

1. Will temperature change make water change?

LEARNING OBJECTIVE: To observe how changes in temperature transform water from liquid to solid and from solid to liquid.

MATERIALS:

Water

Two identical shallow plastic bowls (Do not use ice cube trays. In some parts of the country, children may believe that ice is formed only in cube shapes.)

Access to a freezing compartment or, better yet, to freezing weather

SMALL-GROUP ACTIVITY:

1. Early in the morning, let children fill bowls almost to the top with water. Help them deliver one bowl to the freezer or to the near-zero outdoor location. Leave the other bowl in the room (out of spilling range) for an hour or more.
2. Ask: "Do you think both bowls of water will stay the same or will one change? Let's check."
3. Bring ice back to the room when it is solid. Dump it out onto a pie pan for all to see.
4. "What do you think will happen if we leave the bowl of water in the cold place? Let's find out." Repeat the activity if the children request it.[*]

[*]Some children may have had enough experience with water and ice to predict reversibility of this change in water's form. The transformation from water to vapor through temperature change is less familiar. Both forms of reversibility may need repeating several times for less experienced children to begin to clarify cause-and-effect ideas.

5. Bring snowballs or icicles indoors to melt in a pan, when possible. Collect snow in a coffee can and measure its depth. Measure the depth of the water, after the snow has melted completely. Is it as deep as the snow was?

2. Can water change to vapor; vapor change to water?

LEARNING OBJECTIVE To observe how changes in temperature transform water from liquid to vapor and from vapor to liquid.

MATERIALS:

Foil potpie pan

Clear plastic tumbler

Dark paper or folder

Water, ice cubes

Thermos of boiling water

Magnifying glass

GETTING READY:

Experiment at home. Form a backdrop with dark paper to make the water vapor more visible.

Try to provide more than one setup to avoid crowding observers.

Repeat as needed.

SMALL-GROUP ACTIVITY:

1. Recall the earlier experience in which air temperature helped water change forms. "Let's watch what happens in this cup."
2. Let children examine the cup and pan to be sure they are dry and free of holes. Caution the children about staying back while hot water is poured into the tumbler. "The cup *looks* empty; do you think it is? Is it easy to see through the cup of air now?"
3. Let children add ice to the pan and feel the air temperature just below it. Pour hot water into the cup to the depth of 2" (5cm).
4. Let children feel the warm air temperature in the cup and the cold air temperature under the pan. "We'll watch for changes in the cup for a few minutes."
5. Immediately cover the cup with the pan of ice (Figure 8–1). "What's happening?" (Evaporation is occurring. Tiny bits of water are mixing with air to form water vapor.) "Is the water vapor moving up to the cool air? Let's lift the pan to see what's happening underneath. What does it look like?"
6. Look through the magnifier at the collected droplets. "How could those drops get on a dry pan?" There should be large drops falling before long. Help children recall water bits going into warm air; warm air rising up to the cold air; moist cooling air changing to larger and larger drops; drops falling back down as water again.

Note: Children are likely to describe the collected condensation drops as rain. Using the same analogy, mention that the visible water vapor in the jar is like a fog or a small cloud. To be accurate, condensation occurs in clouds when droplets collect on dust particles to form raindrops. We should say that the drops in the cup formed almost the same way as rain is formed. Be certain that children know that air far from the Earth is cold because it is far from Earth's warmth. Air close to the Earth is warmed by reflected heat from the Earth. (Be sure that children understand that there are no pans of ice in the sky.)

Read *Hide and Seek Fog* by Alvin Tressalt.

Group Experience: Ask the children to breathe slowly into their cupped hands. Does their breath feel warm or cold when it comes from their bodies? Pass out small mirrors or foil pans for children to exhale on. Are these surfaces cool? Ask for predictions about the outcome of breathing on the cool mirror or metal. Find out what happens. Can the children see results? "How does the cloudy place feel? Is it wet or dry? Try it again. Feel the cloudy place, and then feel a place that wasn't breathed on. Why do we feel wetness?"

FIGURE 8–1

CONCEPT: The surface of still water pulls together.

1. Are drops of water curved?

LEARNING OBJECTIVE: To notice the curved form taken by drops of water and the adhesion property of water drops.

MATERIALS:

Water

Small funnel

Small bottles

Medicine droppers or straws

Baster

Waxed paper or formica-
    topped table

Plastic aprons

Popsicle sticks

Sponge to *absorb* spills

GETTING READY:

Fill individual bottles with
    water.

Children like to do this with a
    funnel and small pitcher. Let
    them help, if possible.

SMALL-GROUP ACTIVITY:

1. Give each child a square of waxed paper. "What will happen if a bit of water is squeezed from the dropper onto the paper? Find out. Did it make a flat splash or a rounded drop? Watch closely."
2. Suggest seeing what happens when many drops are made close to each other and when drops are made with the baster.
3. Talk about the outside edges of the drops pulling together tightly to behave *almost* like an invisible "skin." This tightness is not strong enough to stay very curved when many drops join.*
4. Suggest gently dipping the spoon handle or stick into a big drop while watching closely. Notice how the water clings as the spoon is slowly pulled away.

Extension: For a restful rainy day activity, tint water in cups, so drops of different colors can be mixed together on their papers. The action of different colors jumping together is exciting to watch. Add absorption to the experience another time, comparing what happens with different papers: fingerpaint paper, paper towels, coffee filters.

2. Can surface tension hold some weight?

LEARNING OBJECTIVE: To explore the nature of surface tension.

MATERIALS:

Clear glass or plastic jar,
    about 4" (10 cm) in diameter

Water

Syringe-type baster

Plastic aprons

Sponge and newspapers to
    absorb spills

SMALL-GROUP ACTIVITY:

1. Fill one or two jars almost to the top with water.
2. While children watch with heads close to the table, slowly add drops of water from the baster until the jar is full. Ask: "What might happen if more drops are added?" Find out, drop by drop. Did the surface pull more drops into a curve of invisible "skin"?

---

*The word *tension* means tightness, referring here to the tightness of the clustering of water molecules on the surface of the drop.

Bucket for cleanup

Talcum powder, paper, un-
   cooked spaghetti, aluminum
   foil, waxed paper.

3. Hold a piece of spaghetti with thumb and forefinger. Gently rest it on the surface. Now push it
   through the "skin" vertically. What happens? Try
   other materials. (Only dry pieces rest on the
   "skin.")

3. Does soap break water's surface tension?

LEARNING OBJECTIVE: To cause surface tension to dissipate, using soap.

MATERIALS:

Pepper or talcum powder

Pitcher of water

Small jar

Spoon

Slivers of soap

Shallow foil pans

Liquid detergent

Plastic aprons

Newspaper, bucket, and
   sponge for cleanup

GETTING READY:

Pour 1" (2.5 cm) of water in
   foil pans.

Half fill with water.

SMALL-GROUP ACTIVITY:

1. "If we stir pepper (or talcum) and water together
   in this jar, will they mix? Find out."
2. "Will they mix if we gently sprinkle pepper on top
   of the water? Try it with the pans of water."
3. "What could be keeping it on top of the water?"
   (surface tension, the invisible "skin")
4. "Let's see what happens when we touch the surface with a bit of soap."
5. Let the children experiment, changing water as
   needed.

PRIMARY INTEGRATING ACTIVITIES

Math Experiences

1. Provide a set of measuring cups or three sizes of paper milk cartons for casual
   use by children playing with water. Ask how many small containers of water
   it takes to fill a cup or carton this size, and so on.

2. Let the children record the weight of different materials in dry and wet states.
   Weigh a dry sponge on a kitchen scale. Drop it in a bowl of water. Ask children
   for predictions about the weight of the wet sponge. Reweigh the sponge. Let
   the children decide what made the weight difference. Squeeze the water out
   over a dry bowl. Weigh the sponge again. A discussion of the results could
   bring in more than/less than comparisons. "Are 4 ounces more than 2
   ounces?"

3. Measure water absorption by dried beans. Early in the day, begin to soak 1/4
   cup of dried beans in 1 cup of water in one container; 1/4 cup dried beans and
   2 cups of water in another. Drain off, measure, and return the water after 2
   hours, 6 hours, and next morning. How much water did the beans absorb?

Several small groups can do this with different kinds of beans and different amounts of water. Results can be compared and charted.

Music (Resources in Appendix A)

1. Sing "Row, Row, Row Your Boat" with children joining hands with a partner and touching feet to pull back and forth. Ask, "Does water holds boats up? Is water heavy to push?"

2. Sing "The Eency Weency Spider" after the evaporation experiences, making an adaptation in lyrics. Change "and dried up all the rain" to "and evaporated the rain." Look for rain spouts near the school for children to watch after a rain.

3. Sing the old nursery song "One Misty, Moisty Morning." After experimenting with absorption and repellency, discuss with the children whether the old man was dry inside his leather clothes.

4. Sing this song, after experimenting with forms of water, to the tune of "Twinkle, Twinkle, Little Star":

> *Water, water from the stream*
> *When it boils it turns to steam.*
> *Water, water is so nice;*
> *Freeze it cold, it turns to ice.*
> *Cool the steam; warm the ice*
> *It's water again, clear and nice.*

—J. H.

Stories and Resources for Children

ARDLEY, NEIL. *The Science Book of Water.* San Diego: Harcourt Brace Jovanovich, 1991. Attractive color photographs make the more advanced water experiments easy for older children to follow.

BAINS, RAE. *Water.* Mahwah, NJ: Troll, 1985. Portions of this reference, such as the water cycle, can be read aloud to pre-primary children.*

BELL, J. L. *Soap Science.* Reading, MA: Addison-Wesley, 1993. Surface tension is explored and stretched even more with detergent, providing fun with bubble making. Additional experiments for the fascinated older child.*

BERGER, MELVIN. *All About Water.* New York: Scholastic, 1993. The author's crystal clear, focused narrative style makes this an excellent resource for youngsters. Paperback.*

CASSIDY, JOHN. *The Unbelievable Bubble Book.* Palo Alto, CA: Klutz Press, 1987. Good toy stores sell this interesting book with the equipment and precise directions for making giant bubbles. Discusses the paradox of how detergent makes surface tension weaker but "stretchier."

DEVONSHIRE, HILARY. *Water Science Through Art.* New York: Franklin Watts, 1991. This excellent book first defines, then demonstrates certain properties of water as art media: Water flows, is wet, evaporates, can be absorbed, and forms crystals.*

DURANT, PENNY. *Bubblemania!* New York: Avon, 1993. The most complete, least expensive bubble book includes good experiments to enhance understanding of surface tension, and offers a myriad of explorations and purely pleasurable bubble activities. Paperback.

FLACK, MARJORIE. *The Story About Ping* (Seafarer Edition.) New York: Viking Press, 1970. In this classic story, a barrel tied to his back keeps a houseboat boy afloat after he tumbles into the Yangtze River.

GLOVER, DAVID. *Floating and Flying.* New York: Kingfisher Books, 1993. Easy directions for making a water wheel, a soap-powered boat, and a new recipe for bubble blowing.

GOLDIN, AUGUSTA. *Ducks Don't Get Wet.* New York: Harper & Row, 1989. Try the duck's waterproofing system described here.

JENNINGS, TERRY. *Floating and Sinking.* New York: Glouster Press, 1988. Good experiments and a clear text make sense of this popular, but often only vaguely understood classroom experience.*

LOBEL, ARNOLD. *Owl at Home.* New York: Harper & Row, 1975. Owl lets freezing air into his house. His soup freezes in the bowl. Later, when owl starts a new fire in the fireplace, his soup thaws.

MARKLE, SANDRA. *A Rainy Day.* New York: Orchard Books, 1993. Charming illustrations capture the curiosity of a child enjoying the effects of rain. Objects absorb, repel, or dissolve in the rain, and the water cycle is traced in this pleasant story.

MILNE, A. A. *Winnie-the-Pooh.* New York: E. P. Dutton, 1961 (chapter 9, "In Which Pooh Is Entirely Surrounded by Water"). Pooh and Christopher Robin use unorthodox means of floating through flood waters.

REY, H. A. *Curious George Rides a Bike.* New York: Scholastic Book Services, 1973. George forgets to deliver newspapers because he is so absorbed in making newspaper boats and floating his fleet downstream.

ROBBINS, KEN. *The Elements: Water.* New York: Holt, 1994. All of the forms of water are reverently shown in lovely, soft photographs, accompanied by a lyrical text.

SIMON, SEYMOUR. *Soap Bubble Magic.* New York: Lothrop, 1985. After enjoying bubble play, children are skillfully guided to discover the effects of surface tension, air concepts, and optical effects. Children are encouraged at the end of the book to continue to think about what they see and do.*

TAYLOR, BARBARA. *Sink or Swim: The Science of Water.* New York: Random House, 1991. Offers more advanced experiences with floating and sinking, and considers liquids of varied density that float or sink in water.

TAYLOR, KIM. *Water.* New York: Wiley, 1992. Excellent, relevant color photographs by the author add strength and liveliness to the concise text and clear experiments for independent readers and younger children.

TRESSALT, ALVIN. *Hide and Seek Fog.* New York: Morrow, 1988. This Caldecott Honor Book evokes lovely images of fog. Paperback.

ZUBROWSKI, BERNIE. *Bubbles.* Boston: Little, Brown, 1979. Directions given for making wonderful bubbles: bubbles inside of bubbles, giant bubbles, geometric bubbles, and more.

Poems (Resources in Appendix A)

From *Poems Children Will Sit Still For,* read "Dragon Smoke," by Lilian Moore.
From *Now We Are Six,* by A. A. Milne, read "Waiting at the Window."

*Starred references, written at the young child's level of understanding, can help teachers who have minimal backgrounds in science to expand their knowledge base.

Fingerplays

Here is a fingerplay about evaporation:

In soapy water	(scrubbing motion)
I wash my clothes,	
I hang them out to dry.	(pantomime)
The sun it shines,	(hands form circle)
The wind it blows,	(wave arms, sway)
The wetness goes into the sky.	

—J. H.

Art Activities

Ice Cube Painting. Use finger paint and glossy paper, but do not wet the paper as you would to prepare for finger painting. Instead, offer the children ice cubes with which to spread and dilute the paint while making designs.

Tempera Painting. Let children watch or help you mix dry tempera paint with water. When the finished paintings are hung to dry, use the term *evaporate* to explain how this drying occurs.

For more water-in-art ideas, see *Water: Science Through Art,* by Hilary Devonshire.

Dramatic Play

Housekeeping Play. Wash doll clothes. If possible, hang some to dry in the sunshine, some in a shady location, and some indoors. Compare drying times.

Indoor Water Play. Place two or three dishpans of water inside an empty wading pool. Provide plastic or metal pitchers, plastic funnels, and lengths of tubing for play. Mark several plastic baby bottles with tape and numerals to indicate levels to which the bottles can be filled with water.

Outdoor Warm Weather Water Play. Barefooted children can transfer water from buckets to the wading pool, using a plastic hand pump from the automotive supply section of a variety store. Try putting a bucket on a pile of hollow blocks next to the wading pool for experimentation with a garden hose siphon. Read *Curious George Rides a Bicycle,* by H. A. Rey. Then help children make newspaper boats to float in the wading pool, following the techniques George uses in the story.

Bubble Making. Choose some of these ways to have fun with surface tension:

- Use commercial bubble solutions and ring-tipped wands to wave or blow bubbles. Replenish the solution with a concentrated mixture of 1/4 cup of liquid dishwashing detergent (Joy is good) and 1 cup of water. If available, try the commercial set of giant rings and multiple bubble template, or try plastic berry baskets. *Use these outdoors.* Wet, soapy floors are slippery!
- Provide straws and cans of detergent/water solution to blow bubbles *into.* This is easier for children who aren't able to manage blowing bubbles into the air.

- Try using fat drinking straws as pipes for blowing detergent/water solution bubbles into the air. Show the children how to dip the end of the straw into the solution to let a film collect across the bottom of the straw. Suggest holding the straw slightly downward to avoid dripping solution into the mouth.
- Try using soft plastic funnels to blow giant bubbles. (Have several ready and be prepared to wash mouthpieces before sharing.) Put the funnel upside down into a bowl of detergent solution. Gently blow a few bubbles into the bowl to allow a film of solution to coat the inside of the funnel. Lift out the funnel and softly blow a bubble.
- Talk with the children about how the outside surface of the soap film pulled together, almost like a balloon around the air. The soap or detergent mixture makes a more flexible, stretchier "skin" than the water makes alone.

More intriguing bubble experiences are offered in a neat book, *Bubblemania*, by Penny Durant.

Creative Movement

Bubbles. Add a soap bubble movement stimulus to your collection of emergency ideas (i.e., ideas to use when you have a group of children ready to do something that isn't quite ready for the children). Tell the children that you will blow imaginary bubbles to them to catch. "Here's a high one . . . catch this one on your elbow . . . your shoulder . . . your chin. Don't let this one touch the ground . . . catch this one on a fingertip and blow it back to me . . . pretend that your hands are made of soap film. Blow into them until they can't stretch any more and they pop."

Snowmen. Guide the children as they roll imaginary snowmen, rolling slower and slower as the ball gets larger. "Let's make a smaller ball for a chest . . . now a smaller ball for a head . . . lift them into place. Oh! There it goes . . . snowballs this size are very heavy. Now, be the snowman yourself, all curvy and cold and tall. But wait, the sun is beginning to shine and warm the air. Oh! What's happening to your arms and your body?. ." Continue with your suggestions and the children's responses until the snowman is a puddle that will soon evaporate and vanish without a trace.

Trip to the Lake. Take the children on an imaginary trip to a lake in summer where they can pretend to float on their backs and stomachs while the water holds them up. They can invent ways to swim, dive, and row a boat. "Now it's winter and we can't swim, but it's so cold that we can . . . "

Creative Thinking

Invent a story mixup that includes references to water concepts. Illustrate the story with mounted pictures cut from magazines or catalogs. When the concept is referred to, let it be in the form of an obvious misstatement for the children to catch and correct. Perhaps it could be a camping story where "the children could hardly wait to cool off by skating on the lake in the summer sunshine . . . " or "John and Anne filled the boat with heavy rocks to help it float on the water . . . " or "Dad put the teakettle in the icebox to boil the water for soup."

Food Experiences

Children put many water concepts to use when they cook, such as boiling water to dissolve gelatin, melting ice cubes to thicken it, and freezing fruit juice into Popsicles. Cooking rice for lunch offers both water absorption and volume measurement experiences. Popping corn is an exciting edible way to illustrate a property of water: high temperature changes water to steam. Children are surprised and pleased to learn that moisture inside the kernels of corn changes to steam so fast that it pops the kernel open. The steam is very evident in a plastic domed corn popper.

SECONDARY INTEGRATION

Maintaining Concepts

1. Discuss evaporation whenever clothes have to be dried at school—after play in the snow or a fall into a puddle. It's especially reassuring to a child who is worried about staying neat to know that an accidental stain can be washed out, the moisture will evaporate, and the garment will look fine again. When children come to school in raincoats and boots, look for drops of water still on the outside of the garments—evidence that these materials don't absorb water.

2. Examine puddles in the playground after a rain. Using stones, mark the outline of the puddle size when it is at its fullest level. (Make a chalk outline if the puddle is on concrete or asphalt.) This will make it easier for the children to make comparisons from day to day as they watch for changes in the puddle. They will remind you if you should happen to forget the puddle-checking ritual. Those in the Temperate Zone might be able to see the puddle freeze and thaw as well as evaporate. Relate that cold air causes the water to freeze.

Connecting Concepts

Some of the relationships between air and water can be observed in caring for classroom fish and plant life; some can be effectively demonstrated with simple experiments.

1. Children can see evidence of *air in water* when they watch a covered jar of water that has been allowed to stand in the room. Rows of tiny air bubbles will appear on the sides of the jar. Point out that fish must have air to live and that water takes in air. Fish have their own way of getting the air they need from the water. Point out the safety fact that children cannot get air from the water when they swim. They must learn how to breathe how to let the water hold them up, and how to use their arms and legs to move along when they swim.

2. Plant dependence upon water can be seen quite well in thin-leaved plants that have gone without water for a weekend. An avocado plant droops dramatically, but recuperates within an hour or so after a good watering.

3. Air pressure (see Chapter 7, What Is Air?) can be used to empty the aquarium, to drain water from a large waterplay tub, or to provide outdoor waterplay. Make a siphon by completely filling approximately a yard (1 meter) of tubing with water, pinching the ends together to keep air from entering the tube. Place one end of the tube in the water and the other in the bucket below. The water will drain into the bucket, unless air entered the tube. Explain that air presses on the surface of the aquarium water, pushing water up the tube. Read about how liquids are moved from one place to another in *Messing Around with Water Pumps and Siphons* by Bernie Zubrowski.

Family Involvement

There are many opportunities for families to point out examples of water and water/air concepts at home: moisture condensing on mirrors and windows at bath time; steamy kitchen windows when dinner is cooking; and the "smoke" of moisture condensation that emits from clothes dryer vents in cold weather. All of these amplify the classroom experiences.

Parents are the best supervisors of the field trips that extend water learnings in the most concrete ways, especially swimming pool and beach visits. There, children can stride through shallow water to test its weight and substance. With encouraging parents standing by, children can learn that the buoyancy principles work to support their own floating bodies.

RESOURCES

AGLER, LEIGH. *Involving Dissolving*. Berkeley, CA: Lawrence Hall of Science, 1990. Dissolving, evaporating, and crystal activities. Paperback.

AGLER, LEIGH. *Liquid Explorations*. Berkeley, CA: Lawrence Hall of Science, 1990. See above. Explorations of liquid properties are carefully organized for classroom use. Paperback.

BROEKEL, RAY. *Experiments with Water*. Chicago: Childrens Press, 1988. Experiments with temperature changes, capillary action, surface tension, and buoyancy for children ready for additional challenges. (Photographs show children handling equipment that should be used under adult supervision.)

DEVONSHIRE, HILARY. *Water Science Through Art*. New York: Franklin Watts, 1991. This excellent book first defines, then demonstrates certain properties of water as art media: Water flows, is wet, evaporates, can be absorbed, and forms crystals.*

HANN, JUDITH. *How Science Works* (pp. 130–143). Pleasantville, NY: Reader's Digest, 1991. Good background information on the properties of water.

LEVINSON, ELAINE. *Teaching Children About Physical Science* (pp. 146–167). TAB Books, 1994.

WARD, ALAN. *Water and Floating*. New York: Franklin Watts, 1992. Intended for older children, this book provides fresh, practical ways of exploring the properties of water. It can provide interesting challenges for children who are ready for enrichment activites. The clear, concise scientific explanations can be helpful for beginning teachers as well.

9

Weather

"Who turns on the sun? Where does the wind come from?" Early in life children wonder and worry about the weather. Exploring simple concepts about the sun, water cycle, and moving air makes weather something interesting to observe rather than something to endure. Understanding something about the nature of lightning can soften fearfulness about storms. The following concepts are presented in this chapter:

- The sun warms the Earth.
- Changing air temperatures make the wind.
- Evaporation and condensation cause precipitation.
- Raindrops can break up sunlight.
- Weather can be measured.
- Lightning is static electricity.
- Charged electrons make sparks when they jump.

In the activities that follow, children feel the sun's effects, observe the effect of warm-air movement, observe a small cloud in a cup, simulate a rainbow, record the weather, and imitate the movement of the Earth around the sun.

CONCEPT: The sun warms the Earth.

1. What feels warm on a sunny day?

LEARNING OBJECTIVE: To become aware of temperature differences in direct sunlight and shade, and infer the source of the warmth.

Group Activity: Do this activity on a mild, sunny day. You will need sand, water, and four containers of equal size.

1. Early in the day encourage children to share ideas about the day's temperature. How did they decide what to wear? How else are they affected by the temperature? "Why is it warm today? Let's see if we can find out."

2. Let children spread a layer of sand in two of the containers and pour water in the other two. Let them touch the sand to decide if both pans of sand feel about the same temperature; repeat with water. Move outdoors with the containers. Place one sand and one water container in heavy shade; place the other sand and water containers in the sun. Plan to return to these places in an hour. "Do you think both pans of sand will feel the same when we come back?" Return to check the predictions. Which feels warmer? What might have caused the difference? Where do the children feel warmer, in the sunlight or in the shady place where a building or tree blocks the sunlight? (The sun warms our Earth—the land, air, and everything the sunlight reaches.)

3. Move around the school area and have children touch the building, blacktop, and sidewalks on the sunniest side; then repeat on the shadiest side. Which side feels warmer? Feel a car parked in the sunlight, then one parked in the shade. Feel the top of a rock on the ground; turn the rock over to feel the underside. Is it as bright and warm in the shadow of the school building as it is on the sunny side? If the children haven't noticed, comment that the sun gives the Earth both warmth and light. In places where the sun's warmth and light are blocked, it is cooler and darker. Even thick clouds can block the sun's light and warmth somewhat. "Will it be just as warm tonight as it is now in the sunlight? Find out and tell us tomorrow."

4. Return to the sand and water pans in the sunlight. Compare the temperatures. Is one warmer than the other even though both were in the sunlight? Do any children remember a visit to the beach? Which felt warmer on bare feet, the sand or the water? Compare the warmth of the water and the sand with a sunlit grassy area. Help children decide whether some things take up (absorb) more warmth from the sun than others. (The land gets warmer in the sun than water does.)

CONCEPT: Changing air temperatures make the wind.

1. What makes air move as wind?

LEARNING OBJECTIVE: To observe that warm air moves upward.

Introduction: The A. A. Milne poem "Wind on the Hill" makes a nice introduction to this activity. Encourage children to tell how they know the wind is blowing, since wind is invisible. Summarize their ideas with a comment about knowing that wind is blowing because of what it does to things we *can* see.

MATERIALS:

Part 1

Unshaded table lamp (regular 60- or 75-watt bulb)

Strips of crisp tissue paper about 3/4" × 4" (2 cm × 10 cm)

SMALL-GROUP ACTIVITY:

Part 1

1. Children can find out something about how air moves as wind by experimenting with air that is heated.

2. Let children feel the air just *above* the *unlighted* light bulb. Give each child a strip of paper to hold steadily by one end just above the bulb. What happens to the other end of the paper? (A strip of crisp tissue paper will droop down.)

3. Repeat *after* the bulb has been turned on long enough to become hot. (Urge caution near the hot bulb.) How does the air *above* the bulb feel now? What happens to the paper in the warm air?

4. Recall that when air is moving, it pushes against things. We can tell which way the air is moving because of the direction it pushes things. Encourage children to observe which direction the hot air pushes the tip of the paper strip.
   Safety note: "We must *never* allow anything that could burn to touch a hot light bulb."

5. Does the free end droop down or is it being pushed up so that it is straight or moving up somewhat? (Hot air rises.) Ask if children have seen other evidence of hot air rising. Smoke (carbon particles mixed with heated air and gases) goes up chimneys; steam rises from a teakettle on the stove; balloonists heat the air in their balloons to make the balloon rise off the ground.
   Add, "When the sun heats the Earth, the air above the warm parts of the Earth heats up, too. All the air above warm land rises, just like the air above the bulb moves up. The rising air is part of the reason for wind."

Part 2

Clear plastic tumbler

Chilled water

Steaming water

Small, clear prescription vials

Food coloring

GETTING READY:

If water is heated in the classroom, use a safely located electric pot.

Safety Note: *Do not use immersion water-heating coils directly in a cup.* They are too hazardous to use around children.

Part 2

1. "Another thing happens to air that causes it to move as wind. We can see something like it using water instead of air."

2. "Let's pretend that the cold water in this glass is cold air and the hot water I pour in is like hot air." Add two or three drops of food coloring to the vial of hot water to make it distinguishable. (Children can observe the effect better when they are at eye level with the tumbler.)

3. "How is the dyed hot water moving?" Watch for a few minutes without disturbing the tumbler. Repeat with the glass of cold water and a few drops of dyed cold water. What happens? Does the cold dyed water move slowly upward and stay around the top in a layer or does it mix with all the water right away?

Gathering data first-hand

Comment, "Water and air get lighter when they are heated. Cold water and cold air *move under* the heated water or heated air and push the hot air or hot water upward. We saw it happen with the colored water but we can't see the same thing happening outdoors. Hot air is lighter than cold air, so cold air rushes under hot air and pushes it up. When that happens we feel the cold air rushing under the warm air. The air moves fast. This makes wind. *Air is always moving like this somewhere around the Earth.* Warm air moves up and cooler air rushes in under it, pushing against leaves, flags, people, *everything.*"

Encourage children to find out at home what happens when freezer doors are opened in warm, steamy kitchens. Which way does that foggy air move?

Note: In a follow-up discussion, mention that the movement of suddenly heated air and fast rushing cold air is the cause of thunder. When lightning streaks through a cloud, it moves so fast that it heats air around it very quickly. The cold air pushes under the hot air so fast that it makes the loud sound we call thunder. Let children listen to the very small popping noise they can make by drawing their lips over their front teeth, compressing their lips together, then pushing air out as they would to make the sound of a "b." A small popping noise is made by fast moving air from a small space. Thunder is made by huge amounts of fast-moving air.

CONCEPT: Evaporation and condensation cause precipitation.

1. How is rain formed?

LEARNING OBJECTIVE: To observe water vapor form and condense in sun-warmed air.

MATERIALS

Zip-seal plastic bag

Small, clear plastic vial such as a film canis-
ter, or protective caps from a spray-top
product

Tape

Cranberry juice, or water and food coloring

Group Experience: Ask children, "How do you think rain gets up in the air?" Responses
will vary with age and experience with prior evaporation/condensation/precipitation activities (see Chapter 8). Accept all responses tentatively. "Let's make a model to see what
we can find out for ourselves."

Make a demonstration kit by placing a container half-filled with juice or tinted water inside
the plastic bag and seal it shut. Tape the bag to a sunny window. (See Figure 9–1.) "Let's
watch for several days to see what the sunshine does to the liquid and air in this bag."
Children can observe water vapor clinging to the inside of the bag, then forming into a few
drops that eventually slide down to the bottom. Ask children to draw and report what they
observe. Does the bag look different on a cloudy day, or early in the day compared to noon
sunshine?

Are the raindrops the same color as the juice? Leave a small amount of juice out in the
room in a shallow container for a day or so, until all the moisture evaporates. Did air pick
up the solid bits from the juice or just the water in the juice? Does air pick up salt water over
the oceans to make salty rain? Evaporate a small amount of salted water for a few days to
find the answer. Recall with the group that air picks up water as vapor. Sun-warmed air in
the bag picked up just the water part of the juice.

Warm air can hold more vapor than can cool air. Did the water vapor in the bag look
different on a cloudy day? Outdoors this happens all the time, all over the Earth. The sun
warms the air and hurries the evaporation process. Air picks up bits of moisture everywhere. Vapor collects into clouds as it rises high into the sky. Droplets in the clouds collect
into larger drops. Cool air temperatures high above the Earth hurry this condensation
process. Eventually the drops fall to the earth as rain. Then the evaporation and condensation cycle happens again, and again, and again.

2. *Do water drops change in freezing air?*

LEARNING OBJECTIVE: Children will observe water changing to ice and ice changing to water.

MATERIALS:

Medicine dropper (Narrow
tips make more spherical
drops.)

Cookie-baking pan

Aluminum foil

SMALL-GROUP ACTIVITY:

1. "We've seen water vapor in the bag get cold and
   change to drops of water. What do you think might
   happen to the drops if they got freezing cold before
   they fell? Let's find out."
2. Place foil on the baking pan. Carefully squeeze out a
   drop of water onto the foil for each child, spacing

FIGURE 9–1

Water

Freezer or below freezing
weather

drops so they can't touch. Carry the pan to the freezer
or, on a freezing day, to a sheltered spot outdoors.
3. Return to the freezer in 10 minute. Show the frozen
drops to the children. "Is this how they looked before? What happened?" Quickly spoon a frozen
drop into the palm of each child's hand. "What's
happening now?" Later comment that large frozen
raindrops become ice (*sleet*). Frozen water vapor
becomes snowflakes.

CONCEPT: Raindrops can break up sunlight.

1. Can we make a rainbow?

LEARNING OBJECTIVE: To become aware that sunlight and raindrops can make a rainbow.

Introduction: Open a discussion about rainbows and the joy of being lucky enough to see
one, or even a part of one. Comment that rainbows form when sunlight happens to shine
just the right way into air that holds just the right amount of water vapor after a rain. "We
don't often see rainbows, but we can use a mirror and water to send back some sunlight
through the water to make a bit of a rainbow."

MATERIALS:

Shallow baking pan

White cardboard

Small pocket mirror

Small pitcher of water

A sunny location

SMALL-GROUP ACTIVITY:

1. Place the mirror at a 45-degree angle at one end of the
baking pan. Let children take turns holding a piece of
white cardboard near a sunny tabletop or windowsill
while other children take turns trying to reflect a spot
of sunlight from the mirror to the cardboard.
2. When the spot of light hits the cardboard, slowly
pour water into the pan to cover about 2" (5 cm) of
the mirror. Small adjustments to the angle may be
needed until the white light is broken up into the
spectrum ("like a bit of rainbow") on the cardboard.
3. Verify that *it is light refracted through the water that
caused the light to separate into the spectrum colors.*
To do this, hold a hand above the submerged portion of the mirror, blocking the reflection. What
happens to the spectrum reflection?

CONCEPT: Weather can be measured.

1. How does a thermometer work?

LEARNING OBJECTIVE: To become aware that thermometer liquids expand or contract as temperature affects them.

MATERIALS:

Cooking thermometer

Hot water

LARGE-GROUP ACTIVITY:

1. Recall the experience of feeling warm in the
sunshine and cool in the shade. "How else can

Ice cubes

Small container

Outdoor thermometer

we find out how warm the air is?" Show the thermometer. Record the level of the liquid column (the current indoor temperature). Ask for predictions about where the liquid column might move if the thermometer is taken outdoors for a while. Place the thermometer outside and compare the difference after a while.

2. Examine the cooking thermometer. Note that the numbers start with higher readings than the outdoor thermometer, since food is cooked at hotter temperatures than the weather reaches. Put the thermometer into a container of hot water. "Is the liquid going up or down?" Record the final temperature. Ask for predictions of what will occur if ice cubes are added to the water. Let children add them. Check the results. Read thermometers and record new temperatures. Make a plan to check the outdoor thermometer daily. (Thermometer must be in the shade.) Keep a record of your readings and the daily water forecast. Guide children through the thermometer creative movement found on p. 179.

2. How fast is the wind blowing?

LEARNING OBJECTIVE: To become aware that wind speed can vary.

Group Activity: Move the group to the windows. Can children tell if the wind is blowing? How? What do they see? In a group discussion, introduce the idea that the wind, the heat of the sun, and water work together to make the weather. Winds move rain clouds and dry air.

Weather is more interesting to observe when we know many ways to describe and compare it. For generations people have judged the wind's speed by observing its effect on things around them. The following chart is part of the Beaufort scale of wind velocity and provides visual guidelines for determining the wind's speed.

What to Look For	Description	Miles per Hour
Leaves don't move. Smoke rises straight up from chimneys.	Calm	Less than 1 mph
Light flags blow; leaves and twigs move constantly.	Gentle Breeze	8–12 mph
Large branches sway. It's hard to hold an open umbrella.	Strong Breeze	25–31 mph
Whole, large trees move. It is hard to walk against the wind.	High Wind	32–38 mph
Whole trees uprooted.	Full Gale	55–63 mph

End the group discussion with the measures we take to stay safe in storms. Assure children that weather forecasters collect information very carefully so we can be warned about serious storms and find safety when necessary.

Plan to make a daily check at the windows to decide how fast the wind is moving. Record the description the children decide on a wind chart. Children can check the direction of the wind each day by observing the school flag or by holding up a wet finger to see which side is cooled by the wind. Wind is named for the direction it blows *from.* (A north wind pushes the flag south.)

3. Can we tell how much rain falls?

LEARNING OBJECTIVE: To become aware that amounts of rainfall vary.

MATERIALS:

Widemouthed glass or clear plastic jar

Heavy rubber bands

Small plastic ruler

LARGE-GROUP ACTIVITY:

1. Elicit children's ideas about why rain is needed. Discuss how every living thing needs water to survive, and also that lakes, rivers, reservoirs, and underground water stores are filled by rain. Explore the reasons that it is important to many people to know how much rain falls. (Farm children, those living in flood areas, or those living in cities dependent upon reservoirs for water supply may have information to share about rainfall amounts.)

   "Weather forecasters use special measuring equipment to find out how much rain has fallen. We can make something like it. We can put our rain gauge outside in a safe, open place. After each rain we can check to see how much rain fell." (This will be a relative measurement. It will allow comparisons of the amount of rain collected in the jar from day to day.)

2. Use rubber bands to attach the ruler to the jar. Locate a safe, open place where the jar will be undisturbed. Check the level of water in the jar as soon as possible after a rain. Empty the jar. Keep a record of the number of inches (centimeters) of rain that falls for several weeks during your rainy season.

CONCEPT: *Lightning is static electricity.*

Introduction: Let children share what they understand about lightning. Add, "Lightning is a powerful spark of static electricity. It develops when huge quantities of moisture droplets and tiny particles of ice in rainclouds rub together. Every one of those droplets and ice bits has something invisible that gets active this way: electrons. The activity caused by rubbing together is called *charging.* Let's find out more about electrons that are in everything."

Show the children a 2-inch square of plastic wrap, held between your thumb and forefinger. "What do you think might happen when I let go of this plastic?" Find out.

"Let's see if rubbing (friction) might make a difference in what happens to the plastic." Rub the plastic on a piece of wool or nylon fabric. Pick it up again and let go. "Look at this. I have let go, but the plastic hasn't let go this time. Friction did change some things in the plastic, things so very tiny they are invisible. They are *electrons.* Electrons are part of everything—*even part of us.* Friction gets the energy of the electron ready to move. That is called *charging,* or building an electric charge. Often it is also called *static electricity."* (The children may already have many ideas about static electricity.)

Give children their own pieces of plastic to try. A sample of nylon carpeting or garments that children are wearing can be good synthetic fibers to rub the plastic against. "Electrons are part of everything, but we don't notice them until they are charged enough to move. You can have fun building an electric charge to make some little things jump."

1. Can we build up an electric charge to make things move?

LEARNING OBJECTIVE: To observe the effects of electric charges on noncharged materials.

MATERIALS:

Clear, flat plastic boxes

Tissue paper

Combs (plastic or nylon)

Small pieces of cotton thread

Scraps of wool, fur, silk, or nylon

SMALL-GROUP ACTIVITY:

1. Ask children to tear up tissue paper into rice-sized shreds. Put some shreds in the boxes.
2. "Let's rub a box top with a bit of fabric very fast to see what happens to the shreds of paper." (The shreds are attracted to the top. See Figure 9–2.)
3. Gradually add these ideas: "Rubbing two things together (friction) makes invisible bits, called *electrons*, in one thing (the fabric) leave it and jump to the other thing (the box top). The jumping is called a *static electricity charge*. The charge can pull or push other things. The charged box top pulled the paper shreds to it."
4. Suggest turning the boxes over to build a charge on the bottom of the box. What happens?
5. "Now try to build a charge on the comb with the fabric. Hold a thread near it. What happens? Did the static charge pull the thread to the comb?"
6. "Hold the charged comb near your hair. Does the static charge pull the hair toward the comb?"

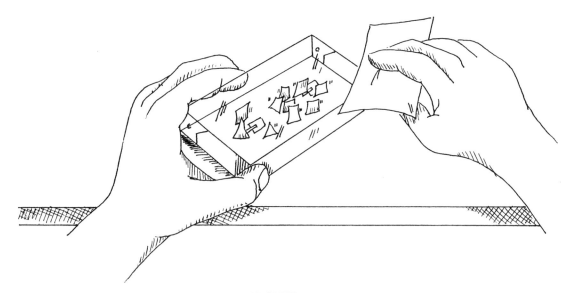

FIGURE 9–2

CONCEPT: **Charged electrons make sparks when they jump.**

1. Can we make a tiny lightning flash?

LEARNING OBJECTIVE: To begin to understand the nature of lightning.

Introduction: "We saw how something charged with static electricity pulled other things to it. Now let's find out if we can see a spark when jumping electrons make a static electricity charge. Perhaps we can see a spark in a dark place. The spark will be like a tiny flash of lightning. We'll pretend that balloons are storm clouds full of water and tiny ice crystals. The water droplets and crystals are blown about and rubbed together in a storm cloud. The energy builds up to make a huge spark: a big charge of electricity."

MATERIALS:

Small, long balloons

Small wool or nylon rug, or carpet sample

Heavy blanket, table

GETTING READY:

Drape table with the blanket to make a very dark place.

Place rug on the floor under the table.

Blow up balloons.

SMALL-GROUP ACTIVITY:

1. Show children how to stroke the balloons on the rug to build the electric charge. Count 20 strokes with them.
2. "Now pretend my finger is another cloud or the earth. I'll bring the charged balloon close to it. Watch! Listen! Did you see the spark jump to my finger? Now you try it."

Group Discussion: "You can make a tiny lightning flash at home in a room that is almost dark. Scuffle your feet on a wool or nylon rug, then touch something metal. You can feel, hear, and see a small charge of electricity jump from yourself to the metal.

"A small spark like this is not dangerous, but lightning from storm clouds is very powerful when it strikes the earth. That is why we take shelter in a building or car during an electrical storm."

Note: The experiences with electric charges will not be successful in a humid atmosphere. Save these experiences for dry weather.

Read age-relevant parts of *Lightning* by Stephen Kramer. Share the awesome photographs.

PRIMARY INTEGRATING ACTIVITIES

Math Experiences

Weather Charting. For the remaining months of the school year, weather data gathering lends itself to chart-making.

For Younger Children: Mark off a calendar on a large sheet of newsprint. Make a yellow sun symbol in each square to remind children that the sun is shining whether we can see it or not. (Speak accurately when asking children about the day's weather: "Can we see the sun today?" not, "Has the sun come out today?")

Cut squares of white tissue paper to tape over the sun on foggy days and opaque white cloud shapes to cover the sun on rainy and snowy days, according to the children's decision about the day's weather.

For Older Children: Record the weather on the calendar by charting the temperature, the class estimate of the wind speed, and sky conditions (clear, cloudy, rainy, sunny). At the end of the month, summarize the total number of days of each condition. A bar graph makes comparisons of the month's weather easy, providing ready answers to questions such as, "Which kind of weather did we have the most often?" and, "How many more days were cloudy than sunny?" Go back to compare monthly or seasonal summaries. Children can learn to identify cloud types and graph the number of days each type was seen. They can watch the weather forecast at home or bring in the newspaper forecast and compare it with the data gathered by the class the next day, recording when the two agree.

A bar graph produces weather records for the group to recall and compare.

Music (Resources in Appendix A)

1. Sing the songs that children have been singing about the weather for generations: "One Misty Moisty Morning," "Rain, Rain, Go Away," and "The North Wind Doth Blow."

2. Sing "The Eency Weency Spider," remembering to change "dried up all the rain" to "evaporated the rain." Check the rain spouts around your building after a rain.

3. Play "Air Cycle Swing" from the Banana Slug String Band cassette, *Adventures on the Air Cycle.*

4. Hang a melodic wind chime near or in your classroom on a breezy day. Let the weather provide the music.

Stories and Resources for Children

ARDLEY, NEIL. *The Science Book of Weather.* San Diego: Harcourt Brace, 1992. Some of the simple projects in this collection can be used as enrichment experiences.

BAUER, CAROLINE (ED.). *Windy Day.* New York: J. B. Lippincott, 1988. This anthology of stories and poems is devoted to windy weather.

BERGER, MELVIN & GILDA. *How's the Weather?* Nashville: Ideals Children's Books, 1993. Written in narrative style for the independent reader, the general weather information is easy to understand. Simple experiments and lively illustrations amplify the concepts. Paperback.

BRANLEY, FRANKLYN. *Flash, Crash, Rumble and Roll.* New York: Thomas Y. Crowell, 1975. Clear information about thunderstorms helps reassure fearful children. Paperback.

BRANLEY, FRANKLYN. *Hurricane Watch.* New York: Thomas Y. Crowell, 1985. Facts are presented about the nature and power of hurricanes, and the protective measures that keep people safe during the storm. Paperback.[*]

BRANLEY, FRANKLYN. *Snow Is Falling.* New York: Harper & Row, 1986. The very readable text describes both the benefits and hazards of snow. A temperature experiment is included. Paperback.

BRANLEY, FRANKLYN. *Tornado Alert.* New York: Harper & Row, 1990. The nature and incredible power of tornadoes will intrigue children. The safety tips will reassure them. Paperback.[*]

DEPAOLA, TOMIE. *The Cloud Book.* New York: Scholastic Book Services, 1974. "If you could hop on a bird and fly way up, you would see . . . " Factual information about cloud types follows. Amusing illustrations. Paperback.

DE WITT, LYNDA. *What Will the Weather Be?* New York: Harper Collins, 1991. Good young child level information about what and how meteorologists study to forecast weather.

DORROS, ARTHUR. *Feel the Wind.* New York: Thomas Y. Crowell, 1989. Clear, simple, and accurate illustrations in a charming text. Instructions for making an easy weather vane are included.

FARNDON, JOHN. *Weather.* New York: Dorling Kindersley, 1992. The brief, well-chosen information in this beautiful book includes simple cloud form identification and easy activities.

GIBBONS, GAIL. *Weather Forecasting.* New York: Simon & Schuster, 1987. A simple description of the tools and work of meteorologists in a weather station tracking the weather of four seasons.

GIBBONS, GAIL. *Weather Words and What They Mean.* New York: Holiday House, 1990. Clearly presented weather definitions.[*]

HISCOCK, BRUCE. *The Big Storm.* New York: Atheneum, 1993. Meteorology concepts are skillfully embedded in context with this fascinating story of one storm as it slowly sweeps from the west coast to the east coast. This mind-stretcher is beautifully illustrated by the author.

KRAMER, STEPHEN. *Lightning.* Minneapolis: Carolrhoda Books, 1991. While some of the explanations of lightning in this book are too advanced for this age group, the narrative style and the safety tips are not. Children will be fascinated by the spectacular photographs.

KRAMER, STEPHEN. *Tornado.* Minneapolis: Carolrhoda Books, 1992. The pace of this book and remarkable photographs match the drama of tornadoes. Explanations and safety tips are provided.

LYON, GEORGE. *Come a Tide.* New York: Orchard Books, 1990. A mountain family copes with familiar torrential rains and the flood that follows. Illustrations fairly drip off the pages.

MARKLE, SANDRA. *A Rainy Day.* New York: Orchard Books, 1993. A lovely story, beautifully illustrated, weaves water and weather concepts into pleasant listening for young children.

MARKLE, SANDRA. *Exploring Summer.* New York: Avon, 1991. This book of summer weather and outdoor activities includes directions for making a simple solar cooker. Also in the series of weather projects and seasonal outdoor activities are *Exploring Winter* and *Exploring Spring.* Paperback.

MINARIK, ELSE. *Little Bear.* New York: Harper & Row, 1957. Chapter 1 provides a good take-off for discussing adaptations to cold weather for young children.

MONCURE, JANE. *What Causes It? A Beginning Book About Weather.* Elgin, IL: The Child's World, 1977. Simple explanations of rain formation, thunder, and rainbows. Attractively illustrated. Good for young children.

POLLACO, PATRICIA. *Thunder Cake.* New York: Philomel Books, 1990. A wise grandmother dispels a child's fear of thunder as they bake together before a storm breaks. Appealingly illustrated.

SABIN, LOUIS. *Weather.* Mahwah, NJ: Troll, 1985. Basic information about weather conditions and forecasting make this a good reference for primary-grade classrooms. Paperback.[*]

SANTREY, LAURENCE. *What Makes the Wind?* Mahwah, NJ: Troll, 1982. This simple, clear book can be read selectively to young preschoolers, and in its entirety to older children. The breezes and gales illustrated in Bert Dodson's watercolors almost lift off the pages. Paperback.[*]

SUZUKI, DAVID. *Looking at Weather.* New York: Wiley, 1991. An enthusiastic presentation written for advanced readers by a TV weatherman, this book includes careful directions for 25 weather experiments. It has directions for making a solar cooker, and for getting a weather report from cricket chirps.[*]

TRESSALT, ALVIN. *Hide and Seek Fog.* New York: Morrow, 1988. The classic Caldecott Award winning book for younger children evokes the feel of fog. Paperback.

WYLER, ROSE. *Raindrops and Rainbows.* New York: Simon & Schuster, 1989. Simple text and interesting experiments, such as measuring the size of raindrops, and exploring aspects of rain, clouds, and rainbows.

Poems (Resources in Appendix A)

A good poem for stimulating weather imagery is "Spring Morning," by A. A. Milne, from *When We Were Very Young.*

From A. A. Milne's *Now We Are Six:* "Wind on the Hill" reveals a child's problem-solving logic about where the wind blows, and the unsolved question of where the wind comes from.

---

[*]Starred references, written at the young child's level of understanding, can help teachers who have minimal backgrounds in science to expand their knowledge base.

From Robert Louis Stevenson's *A Child's Garden of Verses:* "The Wind," in which another child tries to understand the wind; and "Rain," four classic lines of verse.

Lee Bennett Hopkins has compiled readable, likeable poems in his new book, *Weather.*

Art Activities

Weather Mobile. Let children help you create weather symbols to hang in your classroom. Cut out pairs of cloud shapes from newsprint. Staple them together over a wire coat hanger. Let children stuff them with shredded paper or polyester fiber to create a soft sculpture appearance. Place ends of plastic silver tinsel on strips of tape. Fold the tape lengthwise over the bottom of a white wire coat hanger to simulate rain. Let children crayon spectrum colors on a cardboard rainbow arch. Add a yellow construction paper sun to the mobile.

Fog Pictures. Let children draw outdoor scenes on light blue paper. Attach a sheet of white tissue paper of matching size to the top edge to give the effect of fog. The "fog" can be lifted by folding back the tissue paper.

Rain Painting. See what changes will be created by holding children's finger paintings outdoors in a light rain for a few seconds.

Sidewalk Painting. Give children wide brushes and cans of water to paint huge designs on the sidewalk on a cloudy day. Ask for predictions about what will happen to the designs when the sun comes out or a warm wind blows over them.

Good directions are given for making sun prints in *Good Earth Art,* by Mary Ann Kohl and Carolyn Gainer.

Creative Movement

Thermometers: "Find a space where you can stand without touching anyone. Pretend that you are a special liquid that gets bigger when it is warm and smaller when it is cool. Now you are going into a long, skinny tube with a round space at the bottom. Are you standing straight up with stomachs pulled in tight so that you fit into the thermometer tube? I'll tell you when the sun is warming the air around you. When you get warm and grow bigger, you can only go up in your skinny tube. When you shrink and get smaller in cold air, you can only sink straight down.

"Now the sun is shining in a clear blue sky. It is lunchtime. The air is getting hotter and hotter. You have to get bigger and bigger because the air is so warm. Oh! Your fingertips have reached all the way up to 40 degrees C on the number marks. I'll pretend to carry you indoors where it is cooler. The sun isn't shining inside. Ah, it feels cooler here. What will happen inside the tube when you are cooler? There you go shrinking down a bit. It must be about 25 degrees C in here. Now I'll take you outdoors again. It's getting warmer. But wait, a big cloud comes between me and the sun. I can't see the sun now. What happens to you? Do you get bigger or smaller? I see some liquid columns starting to shrink down when the sunlight is blocked and the air feels a bit cooler. Now it is night; the sun isn't shining on our part of the Earth. You are shrinking down more and more because it's cooler than it was in the daytime sunlight. Let's see what happens when a snowstorm happens. Look, there are icicles and snowmen. Now I see liquids shrinking down. Let the liquid slide all the way down to the floor. Now become yourselves again."

Creative Thinking

Print as a story, or tape-record children's responses to stimulus phrases such as "If I could drift on the wind like a bird . . .," "If I were a raindrop . . .," "If I could sail away on a cloud . . ." Read the completed story or play the finished tape for the whole class to enjoy. Make a booklet of the ideas evoked by the children, including illustrations by each idea's author.

Food Experiences

For centuries warm air has been used to preserve fruits, vegetables, fish, meat, and herbs. Children enjoy preparing dried apple rings. Let them wash several apples. Then you core the apples and slice them crosswise into 1/4" (0.6 cm) rings. Children then drop the apple rings into a bowl of water to which 2 tablespoons of lemon juice have been added. Pat rings dry. String them on a long, clean dowel or on a heavy cord. Tie the cord or place the dowel across a warm corner of the classroom, away from direct sunlight, to dry for several days. Apples will be dry to the touch, darker in color, and slightly rubbery. Rinse slices, dry, and eat. Read about apple and berry drying during pioneer days in *The Little House Cookbook* by Barbara Walker.

Experiment with drying small clusters of grapes on a towel-covered tray placed in the sun. Turn grapes each day. Cut fresh corn from the cob. Dry in the same way. Feed the dried corn to the birds, or soak it overnight, simmer for 15 minutes, and serve. Bring samples of dried legumes, dried fruits, or dried beef, if possible.

Field Trips

After a rain stops, tour the school neighborhood to examine the effects of wind and rain on the environment. Where has water collected in a puddle? Why there? Is a wooden sandbox seat as wet as a blacktop area? Is the ground under the playground swings as dry as a graveled area? Have leaves been blown down (small branches, large branches)? What might these signs indicate about the speed of the wind? Was there enough rain to flood the gutters or to wash earthworms out of their tunnels?

SECONDARY INTEGRATION

Maintaining Concepts

1. Expand the weather chart with questions about adaptations children may have made to cope with the day's weather. It can become part of the day's routine to inquire about how the weather affected their lives positively or negatively.

2. If a small dehumidifier is available, examine it on a humid day with the children. See how this adaptation to change a room's climate uses the process of condensing water vapor to water. Unless your dehumidifier poses a special safety hazard, let the children feel the condenser coil before turning on the unit. Switch on the unit and let children hold a piece of paper next to the condenser, then in front of the blower screen, to discover which way air is being moved. Watch the changes gradually taking place on the condenser coils. Does the coil feel the same after the unit has run for a while as it did before? Let children help empty the water from the collection pan. Help them relate this process to the cloud in a jar and to clouds in the cold atmosphere.

Connecting Concepts

1. A miniature water cycle can be observed in a classroom terrarium. At times condensation will be visible inside the glass enclosure. At other times it will appear to be dry. The plants survive without additional watering for many weeks. Why?
2. Recall the experiences with water changing forms; with air taking up water; with the effects of moving air; with water pushing up objects that are less heavy than water; and with air supporting gliders.
3. Note that the water cycle (the evaporation/condensation process) is one of nature's ways of keeping our limited supply of water clean.

Family Involvement

Encourage families to show children the cold, wet air that rolls down from an open freezer door on a humid day, or the direction hot steam takes when it escapes from pots of simmering food or from a teakettle. Suggest that families owning Swedish holiday candle chimes examine the before and after effects of heating the air below the figures with the small candle flames. Note how the blades of the candelabra are bent to catch the rush of hot air and begin to move. Tell families what nearby homes or businesses have visible solar energy panels on their rooftops. They could mention that the sun's energy can be used to heat water in those homes. Mention the location of local windmills used to generate electricity. Suggest that families discuss the things they can do to save energy used to heat and light homes.

RESOURCES

EARTHWORKS GROUP. *50 Simple Things Kids Can Do to Save the Earth.* Kansas City: Andrews & McMeel, 1990. Two weather-related eco-experiments are described, dealing with the effects of smog and acid rain.

LEVENSON, ELAINE. *Teaching Children About Science.* Blue Ridge Summit, PA: Tab Books, 1994. Good ideas are offered for explaining the water cycle, as well as an excellent analogy experience for understanding the expansion of warm air.

MANDELL, MURIEL. *Simple Weather Experiments with Everyday Materials.* New York: Sterling Publishing, 1990. Readable background information on weather phenomenon, with additional experiments to illustrate more advanced concepts.

McVey, Vicki. *Weatherwisdom*. San Francisco: Sierra Club Books, 1991. General background information on weather, plus activity suggestions that could be adapted for Early Childhood classrooms.

Rubin, Louis. *The Weather Wizard's Cloud Book*. New York: Workman Publishing, 1989. "A unique way to predict the weather accurately and easily by reading the clouds."

VanRose, Susanna. *Earth*. London: Dorling Kindersley, 1994. This excellent Eyewitness Series book includes basic background information weather.

Teaching Resource

Children's Television Workshop. *The Big Bird Gets Ready for Hurricanes* kit. Hurricane safety information, planning for a hurricane "watch" and "warning." Includes booklet, game, and audiotape of Big Bird's song. $2.25 for a single copy. Mail check payable to: Children's Television Workshop, Dept. NH, One Lincoln Plaza, New York, NY 10023.

10

Rocks and Minerals

". . . Oooooh! This one has shiny speckles! Look! This one is all little rocks joined into one. These three are a family. This one has a stripe all around and around." Children relish sifting through piles of rocks to find favorites. They are awed to hear that our Earth is a giant ball of rock and that rocks can be millions of years old. Learning about the importance of rocks seems to promote feelings of security in youngsters, just as the Rock of Gilbraltar symbolizes trust and stability to adults. Experiences in this chapter explore the following concepts:

- There are many kinds of rocks.
- Rocks slowly change by wearing away.
- Crumbled rocks and dead plants make soil.
- Old plants and animals left prints in rocks.
- Minerals form crystals.

An unstructured experience in washing and examining ordinary highway gravel is suggested as a beginning activity. This is followed by classifying and hardness testing. Basic information is given about rock formation. Other suggested experiences include grinding soft rocks, breaking rocks open to compare the fresh surfaces with the worn exteriors, pulverizing rock to compare it with complete soil, and forming crystals.

Introduction. Tuck a small rock into your hand, then comment quietly, "I have something quite small but very old in my hand. It is so old that it might have been here long before any people lived on Earth or even before the dinosaurs were alive. Can you guess what it could be?" Slowly open your hand to reveal the humble, but now significant, rock.

CONCEPT: There are many kinds of rocks.

1. How do rocks look when dry and wet?

LEARNING OBJECTIVE: To notice and describe the appearance of a variety of rocks.

MATERIALS:

Bucketful of # 67 washed gravel (builder's supply store)*

Dishpans, water

Few drops of detergent

Old, small brushes

Old newspapers (for under pans)

Trays for clean rocks

Cleanup sponge

Smocks for children

GETTING READY:

Fill dishpans one-fourth full with water.

Note: Children seem to be drawn to a material that is available in quantity for them to explore freely. A heaping dishpan of common rocks will have more appeal than a box of neatly labeled special rocks.

SMALL-GROUP ACTIVITY:

1. Invite the children to enjoy looking at and washing the rocks. "Do wet rocks look the same as they did when dry?" Listen to children's comments.
2. Leave the gravel out for children's independent examination during unstructured times, if possible.

2. Can we find rocks that are alike in some ways?

LEARNING OBJECTIVE: To group rocks into simple classes.

MATERIALS:

Bucketful of # 67 washed gravel*

Sorting containers (egg cartons, margarine tubs, divided plastic snack trays, old muffin tins)

SMALL-GROUP ACTIVITY:

1. Suggest putting rocks that are alike in some way together in containers.
2. Classifying decisions are easy for many children to make (by color, size, shape, texture, pattern). If children can't get started on their own, offer easy ideas such as, "The flat rocks go here; rocks that aren't flat go there."

3. Which rocks are hard; which are soft?

LEARNING OBJECTIVE: To use hardness tests as the basis for grouping rocks.

*# 67 washed gravel consists of varied rocks (from old riverbeds) that range from almond to egg size. For school use, some building suppliers might give you a bucketful without charge. It is usually sold by the ton.

MATERIALS:

#67 washed gravel

Blackboard
  chalk (pressed gypsum)

Pumice (from the drugstore)

Pennies

Three trays or boxes

GETTING READY:

Make two labels: *soft* and *hard*

Place trays next to each other.

Put *soft* and *hard* labels in first
  and third boxes; leave middle
  box unlabeled.

SMALL-GROUP ACTIVITY:

1. "Some rocks are soft enough for a fingernail to
   scratch them. Rocks just a bit harder can't be
   scratched that way, but a penny edge will scratch
   them."
2. Let children experiment with scratch tests.
   Suggest sorting rocks by degree of hardness: "Put
   rocks scratched by your fingernail into the tray
   marked *soft;* put rocks scratched only by the
   penny into the unmarked tray; put rocks that can't
   be scratched by a penny into the tray marked
   hard."

If children find it hard to form three classes of rocks, simplify the experience to form
two classifications of hardness: hard rocks, not scratched by a penny; and softer rocks,
scratched by a penny. Make a separate activity to find soft rocks, scratched by a fingernail,
and harder rocks, not scratched by a fingernail. Perhaps a teacher is wearing the hardest
rock of all on her finger—a diamond (crystallized pure carbon).

ROCK FORMATION

As children engage in the suggested experiences, they can be fed bits of simple information about the mineral content of rocks and about the way rocks were (and
are) formed.

Mineral Content

When children question the specks and streaks of color they see as they sort rocks,
tell them that most rocks are mixtures of many kinds of material. The materials are
called *minerals.* There are about 2,000 different minerals, but only about 20 minerals are abundant. Different mixtures of minerals make different kinds of rocks.
(You might want to recall baking experiences with children. Talk about how sometimes the same materials are put together in different ways to make different things
to eat. Corn bread and cookies are almost alike in ingredients.) Sometimes the mineral mixtures are easy to see as specks, sparkles, and stripes. Rocks may contain
many different combinations of minerals.

Three Types of Rock Formation

Igneous. Some rocks were once mixtures that melted deep inside the Earth.
When they cooled, they formed rocks.

Sedimentary. Some rocks were formed in layers or parts, like a sandwich.
Layers of sand, clay, and gravel were pressed together very hard at the bottom of
old lakes, rivers, and oceans. Sometimes animal shells were mixed into the rock. It

takes thousands of years of very hard pressure to form rocks this way. Some sedi-mentary rocks have stripes of color or line marks in them. Some feel sandy. Some have tiny rocks stuck together. They were all made under pressure.

Metamorphic. Sometimes rocks that were already formed by pressure or by the cooling of melted mixtures were changed again by more heat and pressure, changing the rocks into still different types.

Try to find out something about the rocks in your area, both the visible out-croppings and the rocks beneath the soil. Were rocks heaved up into mountains and hills in your area millions of years ago? Did ancient rivers wear away rocks and carve out valleys? Mention this to the children when a suitable time arises.

CONCEPT: Rocks slowly change by wearing away.

1. Can we wear away bits of rock?

LEARNING OBJECTIVE: To observe and experience one of the ways rocks change.

MATERIALS:

Pieces of soft, crumbly rock
(shale, soft sandstone)

Tin cans with lids (coffee
cans, large tobacco cans, co-coa tins)

SMALL-GROUP ACTIVITY:

1. Put several soft rocks in each can. Ask children to feel the rocks. Cover the cans tightly. Let children hold then tightly and shake them as long and as hard as they can. Open cans. "Has anything changed? Is there dusty stuff at the bottom?"
2. If this must be an indoor activity, try briskly rub-bing two crumbly rocks together over a piece of white paper.

Group Discussion: Talk about the results of the rock-shaking activity in terms of wearing away pieces of rock. Together, look at the edges of the smoothest pebble you can find and at the edges of a freshly broken rock (one you have broken open with a hammer). Encourage children who have visited natural beaches to recall whether they found smooth pebbles or jagged rocks on the sand. Talk of how pieces of rock break from large rocks, how waves tum-ble the rocks against each other and smooth them, and how sand bits are rubbed off. If there are nearby rock formations with which children are familiar, speak of how they were slowly changed and worn by winds or running streams of water.

2. Can we draw with rocks?

LEARNING OBJECTIVE: To experience the idea that bits of rock wear away.

Group Activity: If there is a safe sidewalk area where this activity can be performed, give children assorted soft rocks to use for outdoor sidewalk art. Mention that cave dwellers made pictures on rock walls with soft rocks of different colors.

Later, show the children two other rocks we use for writing: pencil "lead" (graphite) and blackboard chalk (pressed gypsum). They make marks because they are so soft that worn-away bits are left as the drawing or writing we see.

Does your schoolroom have an old-fashioned slate blackboard to draw on? (Slate is a metamorphic rock.) Perhaps you can buy an inexpensive one at the variety store. Slabs of slate can sometimes be found in salvage stores after wrecking crews have taken off old slate roofs. (They are fine to use under flowerpots on the windowsill after the writing experience is over.)

3. Do rocks look different on the inside?

LEARNING OBJECTIVE: To discover firsthand that weathering changes the outside appearance of rocks.

MATERIALS:

Rocks

Hammer

Old jeans fabric (pre-
  caution against flying
  rock chips)

Safety goggles

SMALL-GROUP ACTIVITY:

1. All participants should wear safety goggles. Cover a rock with the fabric. Strike it hard with the hammer. (Children experienced with hammers can do this successfully. A two-handed grip may help.) Very hard, round rocks may not split well. Other rocks will split with one hammer blow.
2. Examine the inside; compare it with the outside appearance.

Note: This is a very exciting activity that may lead to "rock fever." Children will be curious to discover what is inside a rock: the glint of mica flakes, a streak of mineral, a sleek surface. Engage children's imagination by stating that the worn outside surface of the rock was once the same color and texture as the inside. Ask: "What might have happened to change the outside of these rocks?"

If there is access to a rock tumbling machine and appropriate school permission, the class can embark on an intriguing 6-week rock-polishing project. This will help them understand how slowly rocks change. Each stone will be transformed into a smooth, gleaming gem, perhaps one for each child to have. The machine can be placed in a covered, inexpensive foam picnic cooler to muffle the noise. Children can appreciate the changes in appearance every 2 weeks when the grit size is changed. (See Teaching Resources to purchase a machine by mail.)

CONCEPT: Crumbled rocks and dead plants make soil.

1. What happens when we pound soft rocks?

LEARNING OBJECTIVE: To become aware that rocks are part of soil formation.

MATERIALS:

Pieces of crumbly rock (shale,
  soft sandstone, etc.)

Pair of old, discardable jeans

Hammers

Newspapers

Sieve

Empty can

SMALL-GROUP ACTIVITY:

1. Put one jeans leg inside another. Put a few rocks in the double jeans leg and place on sidewalk. Let children take turns pounding rocks, checking results often.
2. Spread newspaper on the ground. Fit sieve over the can. Shake contents of bags into sieve, catching spills on newspaper.
3. Keep pulverizing until children are tired.
4. Examine the powdered rock in the can and save for the next experience.

2. What does soil look like?

LEARNING OBJECTIVE: To observe that many materials make up fertile soil.

MATERIALS:

Trowel

Container

Newspapers

Topsoil

Sieve, sticks

Magnifying glasses

Container of pulverized rock
  from previous experience

2 paper cups

Water

SMALL-GROUP ACTIVITY:

1. Let children scoop up a trowelful of soil from a permitted digging area to bring indoors.
2. Spread the soil on newspapers; examine with magnifying glass. Compare with rock pulverized by children.
3. Put soil through sieve. Good topsoil contains bits of leaves, twigs, roots, and worms. Some of this matter will be left in the sieve.
4. Put some sifted soil and powdered rock in separate cups. Stir a bit of water into each. Compare.

Note: Clay is like the moist powdered rock. It must have lots of old vegetable and animal matter (and perhaps sand) added to it to make good soil for growing things.

CONCEPT: Old plants and animals left prints in rocks.

Recall splitting rocks open to find unexpected color, sparkle, and texture inside. Add that prints of animals, shells, and plants that lived on Earth millions of years before people did can be found pressed between layers of rocks. These rocks are called *fossils*. (Perhaps there are some fossils in the bucket of highway gravel that the children have been using.)

Children can have fun making prints of their hands, shells, or leaves using the kind of powdered stone that became clay. They can make prints similar to fossils, though they won't be as old or as hard as fossils, of course.

1. Can we make a "fossil" print?

LEARNING OBJECTIVE: To become aware of one way that fossils are formed.

Let children use small rolling pins to roll out moist clay slabs about 1/4" (1 cm) thick. Show them how to press a leaf, shell, or their hand firmly into the clay. It may take several tries to get a clear print. "Erase" unsatisfactory leaf prints by rubbing the clay with water and lightly rerolling. Cut the slab into a plaque shape by using an empty coffee can as a cutter. Poke a hole through the top of the slab for a hanging loop of string. Scratch the child's initials in the clay. Allow several drying days. Mention that fossils took even longer to form. Teacher may shellac the dried clay to preserve it, if desired.
    Read *Dragon in the Rocks* by Marie Day.

CONCEPT: **Minerals form crystals.**

Introduction: (Do this after children understand evaporation.) Look together at some mineral crystals: a diamond or other precious stone jewelry, the rocks you cracked open to reveal glittering bits of mica or quartz crystals, or some large or small salt crystals. Say that when minerals are dissolved in liquid which then evaporates, or when minerals melt and cool slowly, they form *crystals.* Each type of mineral forms into its own special crystal pattern. "We can try to dissolve a mineral in liquid that will evaporate to leave pretty, powdery crystals."

1. *Will crystals form on coal?*

LEARNING OBJECTIVE: To observe the formation of crystals.

MATERIALS:

Coal, broken into small chunks (charcoal briquettes may be used)

Glass pie pan or low ceramic bowl

Old tablespoon

Pint jar for mixing

4 tablespoons *plain* salt

4 tablespoons water

2 tablespoons *clear* household ammonia*

4 tablespoons laundry bluing

Food coloring in squeeze bottles

GETTING READY:

Spread newspapers on work area.

Have smocks for children

SMALL-GROUP ACTIVITY:

1. Heap chunks of coal in pie pan.
2. Let children take turns measuring salt and water.
3. Tell children to stand away from table when ammonia is opened and to hold their noses. An adult should measure the ammonia and bluing (it stains). Mix all ingredients in jar until salt dissolves.
4. Slowly pour mixture over coal pieces to saturate. Let children sprinkle drops of food coloring in separate areas on the coal.
5. Place dish where it can be observed but not disturbed.

Crystals may begin to form on the coal within a few hours and continue to form for several days. The dissolved minerals move to the surface of the coal as the liquid evaporates and the mineral residue forms crystals. Liquid will continue to move to the surface of the newly formed crystals to form crystals on top of crystals. Some children may think they are growing as a plant grows. Point out that minerals can change, but only living things grow.

2. *Will crystals form from salty water?*

LEARNING OBJECTIVE: To observe the formation of crystals.

Safety Precaution: Keep ammonia bottle tightly capped until needed. Open well away from the face. *Keep ammonia out of the possible reach of children.*

MATERIALS:

2 tablespoons rock salt
  (halite)

1 cup hot water

Glass pint jar

String

Pencil

SMALL-GROUP ACTIVITY:

1. Pour hot water into the pint jar. Let children measure rock salt into water, stirring until dissolved.
2. Tie string around the pencil. Place it across the jar top so the string dangles in the salt solution.
3. Place jar on a sunny windowsill. Check the next day for beginning crystal formation. Watch for several days. Use a hand lens to examine the crystals. If you have access to a simple microscope (40x will do), carefully transfer the strand of crystals to a slide to examine the beautiful structure of the crystals.

PRIMARY INTEGRATING ACTIVITIES

Math Experiences

Counting. The Grab Bag Game may be played with stones. Place a container of stones in the center of a circle of children. Let them take turns reaching in to scoop out as many stones as they can hold in two hands. Then count the results. Keep score with an abacus or a hand-held calculator. Use a numbered spinner, or make dice from 4" (10 cm) squares of cardboard to determine how many rocks players win.

Buried Treasure. If a sandbox is available, bury a specific number of stones (or stones and shells, if you wish to include classifying) for children to hunt. A large sandbox may be "staked out" with string to form separate territories for several children to use amicably.

Counting Cans. Mark numerals from 1 to 10 (or more) on the sides of low metal cans. Line up the cans in order on a table or shelf. Place a coffee can full of gravel beside them for independent counting of appropriate quantities of stones to put in numbered cans.

Ordering. Search through the bucket of gravel to find stones of conspicuously different sizes. Place them in a paper bag. One child at a time can reach into the bag to choose a stone to order by size. Allow time for much fingering and feeling in choosing. Ask for the smallest stone found. Next ask children to compare their stones with the smallest one to find a stone just a bit larger to put next to it. Build up to the largest stone. Older children who have had experience measuring with rulers could use them for this activity. Suggest ordering rocks according to texture, from smoothest to roughest; and color, from lightest to darkest.

Weighing. Keep two coffee cans full of stones near the classroom balance for independent use by the children.

Music (References in Appendix A)

1. A familiar folk song can be easily adapted to help children remember that mountains are rock formations. Sing "The Bear Went Over the Mountain," substituting "... But all that he could see was the rocky part of the mountain" for "... But all that he could see was the other part of the mountain." You could mention that some mountains are covered with soil and growing things while other mountains are bare rock on top.

2. Hear Michael Mish sing, "Center of the Earth" on his cassette, *I'm Blue.*

Stories and References About Rocks and Minerals

ALIKI. *Fossils Tell of Long Ago.* New York: Harper Collins, 1990. A cheerful presentation of fossil formation and how they inform about the ancient past. Paperback.

BAYLOR, BYRD. *Everybody Needs a Rock.* New York: Scribner's, 1985. A young child's view of rock hunting.

BAYLOR, BYRD. *If You Are a Hunter of Fossils.* New York: Scribner's, 1984. Lovely fragments of prose evoke images of the ancient plants and animals found as fossils.

BRANLEY, FRANKLYN. *Earthquakes.* New York: Harper & Row, 1990. Children will be impressed by the nature and power of earthquakes. They will be reassured by learning what they can do to take care of themselves, should they experience an earthquake.

BRANLEY, FRANKLYN. *Volcanoes.* New York: Harper & Row, 1985. Calming assurance that good warning systems keep people safe from volcano danger follows the information about this natural phenomenon.

COBB, VICKI. *Lots of Rot.* New York: J. B. Lippincott, 1981. Interesting experiments and lighthearted illustrations demonstrate the important role of decomposers in soil formation.

COLE, JOANNA. *The Magic School Bus Inside the Earth.* New York: Scholastic, 1988. The author and her illustrator, Bruce Degen, entertain delightfully as they present information about Earth's interior.

CURRAN, EILEEN. *Mountains and Volcanoes.* Mahwah, NJ: Troll, 1985. Lovely watercolors illustrate this very simple text that tells readers how slowly mountains change, and how rarely volcanoes erupt. Paperback.

DAY, MARIE. *Dragon in the Rocks.* Buffalo, NY: Firefly Books, 1991. The true story of a persistant girl of 12 who persevered to singlehandedly unearth the rare fossil of a giant fishlizard, the ichthyosaur. It can still be seen over two hundred years later in the Natural History Museum in London.

FARENDON, JOHN. *How the Earth Works.* Pleasantville, NY: Reader's Digest, 1992. This book provides good information on volcanoes, earthquakes, soil formation, and rock identification.

FOWLER, ALLAN. *It Could Still Be a Rock.* Chicago: Childrens Press, 1993. A variety of simple concepts are briefly presented in this small, pleasant book for beginners.

GANS, ROMA. *Rock Collecting.* New York: Harper & Row, 1984. Children are drawn into this book that says, "The oldest things you can collect are rocks ... millions and millions of years old." Basic rock formation and cheerful illustrations are included. Paperback.

GIBBONS, GAIL. *The Pottery Place.* San Diego, CA: Harcourt Brace Jovanovich, 1987. A young friend watches a potter as she mixes clay, throws pots on her wheel, fires and glazes her wares. Three methods are described for children to make simple pottery, a fine extension for rock study.

HISCOCK, BRUCE. *The Big Rock.* New York: Atheneum, 1988. This tale of an ancient rock leads children who scramble on it toward appreciating the constant, slow changing of the Earth's surface.

HUGHES, SHIRLEY. *Alfie Out of Doors Storybook.* New York: Lothrop, Lee & Shepard, 1992. "Bonting" is the story of Alfie's best smooth, gray stone that was found, lost, and miraculously restored to its owner. Charmingly illustrated by the author.

KAUFMAN, JOE. *Joe Kaufman's Big Book About Earth and Space.* New York: Golden Books, 1987. This book compiles capsules of child-level information about the geology, geography, and weather of our planet Earth. A good classroom reference with basic answers to children's questions about earthquakes and volcanoes, along with facts about rocks, minerals, and crystals.

MCNULTY, FAITH. *How to Dig a Hole to the Other Side of the Earth.* New York: Harper & Row, 1990. Earth science facts blend easily into this fantasy journey below the crust of the Earth. Paperback.

O'DONOGHUE, MICHAEL. *Rocks and Minerals.* San Diego: Thunder Bay Press, 1994. Excellent color photographs for rock identification, rock hunting rules and safety tips, and 12 interesting activities make this a good resource for children.[*]

PETERS, LISA. *The Sun, the Wind, and the Rain.* New York: Holt, 1988. This fine lesson in geology concepts parallels, page by page, the forming and changing of a mountain with a child's construction on a nearby beach of a sand mountain. Read this story aloud at the sand table or sandbox as you build and change a sand mountain along with the child.

SANTREY, LAURENCE. *Earthquakes and Volcanoes.* Mahwah, NJ: Troll, 1985. Information for primary-grade children about these powerful natural events will reassure some, and raise concerns for others who live close to the Pacific plate. Paperback.[*]

WYLER, ROSE, & AMES, GERALD. *Science Fun with Mud and Dirt.* New York: Simon & Schuster, 1986. Among the simple activities Wyler describes as "mudpie science" are directions for making a miniature mud house and brickmaking. Paperback.

Poems (References in Appendix A)

Read the profound poem "Rocks," by Florence Parry Heide, in *Poems Children Will Sit Still For.*

Read the following poem with a box of your favorite rocks at hand:

My Rocks

Here in this box
I keep my rocks.
They came from under the ground.
Some are striped, some are plain,
Some are tiny as a grain.
My favorite is round!

—J. H.

Art Activities

Rock Sculpture. Let children create additive sculpture by joining rocks of assorted sizes, using white glue. The resulting sculpture forms may be painted with tempera. Sculptures that have been coated with shellac (adult work) can be used as paperweights. Older children enjoy trying to create fantasy by gluing dyed aquarium stones, seeds, and tiny shells to their sculptures.

Rock or Sand Collage. Small stones, painted if desired, can be glued in patterns onto heavy cardboard. Color sand by shaking it in a tightly covered jar with

[*]Starred referencs, written at the young child's level of understanding, can also help teachers with minimal backgrounds in science to expand their knowledge base.

dry tempera paint. Punch nail holes in the lids to make sand shakers. Let the children paint swirls of diluted white glue on construction paper, using cotton swabs, then sprinkle colored sand over the glue patterns. Surplus sand can be tipped onto a small tray and reused.

Chalk Drawings. When children use chalk as an art medium, recall that the chalk material was once a rock.

Sand Painting. Prepare four colors of sand as you did for the collages above. Give children each a clear, screw-topped container. Let them spoon successive layers of sand into the container, being careful not to disturb the container. Probe gently along the inside of the container with a toothpick to produce interesting designs. My student, Peggy Robinson, uses these layers to help children understand how the time when dinosaurs lived is determined. Add one layer each day, recording the date and color of sand. Place a small chicken bone on one layer. After 5 days ask, "Which is the oldest layer of sand? Which is the newest? How do you know?" Explain that the Earth's crust is made up of layers of rock that have built up on top of each other. Scientists can tell how old dinosaur remains are by the layer of rock in which they are found.

Clay. Remind children of its origin when using natural clay for art projects. Also, refer to Rose Wyler's book, *Science Fun with Mud and Dirt,* if making a model mud hut would extend a social studies concept.

Dramatic Play

Introduce rocks as a play material in the indoor or outdoor sandbox. Rocks can be formed in lines to make the floor plan of a house. They can also be piled up to make furniture, and sticks or clothespins can become the people who live in the house. Lines of rocks can also outline highways for toy cars to travel through sand-table dunes and deserts. In addition, children find relaxed enjoyment using sifters to separate the rocks from the sand at cleanup time.

Creative Movement

Take a "rock walk." Children enjoy moving in a circle, interpreting movement suggestions, and following the rhythm of the teacher's drumbeats or clapping. Adult participation in the activity encourages the involvement of children. Suggest ways to move in a story form such as this: "Let's go for a rock walk. Let's pretend to be barefooted. We'll start on this gravel path. Oh! It's *hard* to put our feet down on the broken rocks. We'll have to move quickly and lightly to the end of the path. Step, step, step, step. There, that's better. We've come to a sandy place. Let's scuffle our feet in the sand. Scuffle, scuffle, push up little piles of sand with each step. Oh look, we've come to a wet place where we have to walk on moss-covered rocks. They are so slippery to walk on; let's move very carefully with each step. Look at that huge rock. Let's try to climb it, bending way over to hold on with our hands. Climb slowly, slowly. Now we're at the top. Let's run down the other side to the beach ahead. Run, run, run to the sandy beach. But this sand is *hot* from the sunshine. We can't move slowly here. Let's move quickly down to the edge of the water to cool our feet. Now let's sit down to rest."

Creative Thinking

As a group discussion with younger children, or as a story-writing theme for older children, imagine what everyday life would be like if nobody ever thought of moving, breaking, grinding or building with rocks.

Food Experiences

Children are surprised to find that we eat small amounts of one kind of rock every day (salt). Show them some rock salt if available. Mention that all the salt that is mined under the ground was once part of ancient oceans.

Another link between food and rocks is the use of grainy-surface rocks to grind seeds into meal and grain. Perhaps you can find a bag of stone-ground flour or cornmeal in a natural foods shop. (One teacher let fascinated children grind dried corn between two rocks as part of a study of American Indians.)

Janice VanCleave suggests making a "sedimentary sandwich" to demonstrate how sedimentary rocks are formed over the centuries, as layers of materials form under pressure to join other layers of rocks. A slice of bread, covered by a layer of peanut butter, then by a layer of jelly, then by another layer of bread is pressed together and cut into slices so the layers can be seen, and then, fortunately, eaten. (From *Earth Science for Every Kid*. See Resources.)

Field Trips

Walk around the school neighborhood to look for the following: rocks in their natural setting, rocks that have been cut for use in buildings, and products made from rocks and minerals. Look closely at the natural rock for signs of weathering, such as cracks or surfaces worn smooth by years of exposure to heat, cold, wind, and rain. Young children personalize familiar rocks according to the events they associate with them: our picnic rock, our sitting-on rock, our storytime rock, and our where-we-found-the-turtle rock. Use the special rock names when you can.

Look for old stone steps or stone curbings that have been worn down into sloping shapes by thousands of feet treading upon them. Look for monuments, stone walls, stone windowsills in old brick buildings, and flagstone or crushed limestone paths.

Man-made materials containing rocks and minerals are everywhere. Concrete or blacktop playground surfacings contain coarse or fine gravel bonded together with concrete or asphalt, both derived from rocks. Cement blocks, bricks, tiles, terrazzo flooring, porcelain sinks, glass, iron railings, steel slides, and fences around school playgrounds are made with rock or mineral raw materials. No school is without structures of this sort to visit. Talking about these strong, solid parts of their environment seems to contribute to children's feelings of security.

Make an ecological scavenger hunt the focus of a field activity, as Margaret Drysdale does. She takes her classes to an empty lot where they check the area for debris. They classify their collected materials into two piles: man-made materials and natural materials. She explains that only the things from nature can be left on the ground, because they will decompose and eventually become part of the soil. The class carries the man-made materials back to the school recycling/trash bin, because those things won't decompose. She emphasizes that a discarded aluminum can or plastic item could still be littering the land when the children are old enough to be grandparents.

SECONDARY INTEGRATION

Maintaining Concepts

Since rocks and minerals are so commonplace, the topic would become tedious if reference were made to them at every opportunity. Do mention rocks when something a bit different prompts a comment, such as, "Is Lynn wearing a very special rock in her bracelet today? Tell us about it." Bring interesting rocks to share with the children whenever you come across one. Talk about its texture, color, or whatever appeals to you. The children will respond similarly with their favorites when they know of your appreciation of rocks.

Connecting Concepts

The inclusion of rocks as a material for experimentation is suggested for the topics of water, and the effects of gravity. It could be pointed out in discussing the effects of magnetism that the original source of magnetic material is a rock called *lodestone* (magnetite). Magnetite is found in this country near Magnet Cove, Arkansas.

The tie between soil formation and growing things is an easy relationship to mention. Wet sand is sometimes used as a growing medium for rooting stems such as begonias or pineapple tops. The powdered rock made when children pounded soft rocks could be used in a seed germinating experience. Compare the results with seeds growing in good topsoil containing humus and other rotted organic matter.

Discuss how thoroughly rotted natural materials improve the soil, so that healthy, strong plants will grow. Ask: "Do people use some plants for food?" Talk about renewing the soil this way as one of the wonderful cycles of nature: from living plants to decomposition . . . from enriched soil to living plants and food again . . . and again . . . and again. Talk about saving and improving the soil with compost as one way that people help to renew our planet.

If possible, in early fall bury several non-biodegradable materials, such as a plastic spoon, a foam cup, or an aluminum can in a marked location on the school grounds. Also include a paper bag and a regular plastic bag. Dig them up in late spring to see if the biodegradable and non-biodegradable materials were affected differently in the soil. Encourage the children to make the connection between what they observed and our need to reuse and recycle man-made objects that do not decompose. Celebrate Earth Day, April 22.

Family Involvement

Invite families to share special rocks or fossils with the class. Masking tape name labels are helpful for insuring the safe return of borrowed rocks. Suggest that they point out to their children areas of rock exposed by highway construction or special rocks that are landmarks in the area. Win the hearts of parents by forewarning them to check the pockets of their children's jackets and jeans for rocks before laundering them, now that the children's interest in rocks has been whetted.

RESOURCES

HORENSTEIN, SIDNEY. *Rocks Tell Stories.* Brookfield, CT: Millbrook Press, 1993. The level of concise information in this book for older children makes good background preparation for teachers of young children.

MEYER, CAROLYN. *Rock Tumbling: From Stones to Gems to Jewelry.* New York: Morrow, 1975. This book may still be on your library's shelves. You will need its complete directions for the rock tumbling process if your polisher lacks manufacturer's instructions.

PELLANT, CHRIS. *Rocks and Minerals.* New York: Dorling Kindersley, 1992. The excellent color photographs in this Eyewitness Handbook make it possible to identify any rock brought in by children, but the descriptions are highly technical.

VAN CLEAVE, JANICE. *Earth Science for Every Kid.* New York: Wiley, 1991. A broad array of Earth science and weather concepts are explored with experiments intended for independent readers.

VANROSE, SUSANNA. *Earth.* London: Dorling Kindersley, 1994. This Eyewitness Science book combines fine illustrations with concise nuggets of basic geological information about rock formation, earthquakes, volcanoes, erosion and more, to make this an intriguing reference for background reading.

ZIM, HERBERT, & SHAFFER, PAUL. *Rocks and Minerals.* New York: Golden Books, 1989. Comprehensive pocket guide for rock identification. Paperback.

Additional Resources:

CHILDREN'S TELEVISION WORKSHOP. *The Big Bird Gets Ready for Earthquakes* kit. Earthquake safety information including a board game and an audiocassette of a song, "Beatin' the Quake." $2.25 for a single copy. Check payable to: Children's Television Workshop, Dept. NH, One Lincoln Plaza, New York, NY 10023.

FEDERAL EMERGENCY MANAGEMENT AGENCY. *Coping with Children's Reactions to Earthquakes and Other Disasters.* FEMA #48. Single copies free. Write to: FEMA, P.O. Box 70274, Washington, DC 20024.

NATIONAL SCIENCE TEACHERS ASSOCIATION. *Earthquake Curriculum.* Federal Emergency Management Agency, 1988. This extensive K–6 multidisciplinary curriculum actively involves children in learning about the nature of earthquakes, recognizing an earthquake, and earthquake safety and survival. Free of charge by writing to: Susan McCabe, FEMA SL-NT-EN, 500 C Street SW, Washington, DC 20472.

Rock Tumbler can be ordered by mail from: Troll Learn and Play, 100 Corporate Drive, Mahwah, New Jersey, 07430. (1-800-247-6106)

11

The Effects of Magnetism

Magnets attract iron, steel, *and the attention of young children.* While magnetic attraction cannot be seen or felt, *its effects can be seen and felt.* Children can accept the reality of an invisible force when they can have experience putting that force to work. The following concepts underlie the experiences in this chapter:

- Magnets pull some things, but not others.
- Magnets pull through some materials.
- One magnet can be used to make another magnet.
- Magnets are strongest at each end.
- Each end of a magnet acts differently.

In this chapter children will experiment with familiar objects to see what magnets will pull and what they will pull through, make temporary magnets, and discover how opposite magnetic poles affect each other. For all magnet experiences, teachers are **cautioned** that computers, discs, and peripherals rely on magnetic force to operate. Children experimenting with magnets near these objects can seriously disrupt them.

CONCEPT: Magnets pull some things, but not others.

1. What will magnets pull?

LEARNING OBJECTIVE: To experience visually and tactually the effects of magnetic attraction and to apply the effects to sort iron or steel objects from nonferrous objects.

MATERIALS:

Magnets of assorted shapes and sizes

Small foam meat tray filled with test items for each child (Iron/steel suggestions: keys, key chains, bolts, screws, nails, paper clips, lipstick cases;

SMALL-GROUP ACTIVITY:

1. Give each child a paper clip and a small magnet to try out. To see the effect well, slide the magnet toward the clip on the table top. To *feel* the effect, hold clip in the palm of the hand.
2. "Do you think the magnet will pull everything to it? Let's find out with the things on the small trays."
3. After some exploration, suggest sorting objects pulled by the magnets from those that are not. Use the *yes/no* trays.

Noniron/steel items: pennies, brass fasteners, rubber bands, plastic, glass, wood, aluminum objects)

Two large trays or box covers

GETTING READY:

Have only magnets on the table when children gather. Count them together. (Tiny magnets are easily misplaced.)

Prepare a mixed collection of objects for each pair of children.

Tape a *yes* label on one tray and a *no* label on the other.

4. "Things on the *yes* tray do not look alike, but they are made of the same stuff." Children may note that all items are metal. You can use the specific terms *iron* or *steel:* "Magnets pull only on the iron or steel objects." (Magnets also attract cobalt and nickel, two minerals not commonly used alone in manufactured articles.)

Note: Toy magnets are rarely strong enough for school use. School suppliers and scientific equipment suppliers sell sturdy magnets. An electric motor repair shop may be willing to give you an old magnet removed from a motor. (**Warning:** Remove your watch before handling a very strong magnet of this size.) Another reusable magnet source is old sound system speakers. Scientific equipment outlet stores carry as many as 40 shapes (rings, cylinders, bars, horseshoes) and kinds (steel, ceramic, rubber) of magnets. (See appendix A for catalogs.)

CONCEPT: **Magnets pull through some materials.**

1. Can magnets pull through things that are not attracted?

LEARNING OBJECTIVE: To discover forms of nonferrous materials through which magnetism will pass.

MATERIALS:

Magnets*

Steel wool pad

Iron and steel objects (nails, washers, bolts, clips)

Paper, cardboard, aluminum

Shoe box

Drinking glasses

Sand or dirt

Water

SMALL-GROUP ACTIVITY:

1. "Do you know what this (steel wool) is made of? Could a magnet help you find the answer?"
2. Tear off bits of steel wool. Give some to each child. Find out if a magnet can pick up steel that is covered with a piece of paper. Try cardboard and aluminum. "Will magnetism pull through these things to pick up steel?"

*The thickness of the noniron material and the strength of the magnet are factors in the success of this experiment. A strong, new magnet will attract a paper clip through one's fingertip. A weak magnet will not attract a paper clip through cardboard. To keep magnets strong, always store two bar magnets together with opposite poles touching (north to south and south to north) when not in use.

GETTING READY:

Put steel object in tumbler.

Put steel objects in box; cover with sand.

Put water in another tumbler; drop a washer in it.

Put water in a third tumbler; drop a washer in it.

3. Touch the magnet to the outside of a dry tumbler. Does it attract the steel object inside the tumbler? Try different magnets and different objects.
4. Suggest dipping magnet into the sand-filled box and the tumbler of water. Can magnetism pass through materials that it does not attract?

Group Discussion: When the children try out various magnets, they will find that some work better than others. Talk about this with the whole group. Children can help keep magnets strong by remembering to put a steel "keeper bar" across both ends of a horseshoe magnet. They can also keep magnets strong by not jarring them.

Michelle delights in the iron filing patterns she creates as her big magnet attracts them through cardboard.

Note: A child may believe that a keeper bar is also a magnet—one that just doesn't happen to work right when it is separated from the magnet. From the child's experience, a plain bar of metal has no assigned function. Find a familiar iron or steel object to use as a temporary keeper bar. "Children's scissors are not magnets. Will a magnet attract them? If so, then we can leave this pair of scissors across the ends of the horseshoe magnet to keep it strong, while you experiment with the other keeper bar." Experience may clarify a child's ideas about the keeper bar, while verbal persuasion from an adult may not.

CONCEPT: **One magnet can be used to make another magnet.**

1. Can we make a magnet?

LEARNING OBJECTIVE: To learn how to make a temporary magnet.

MATERIALS:

2″ (5 cm) needles or straight-
  ened paper clips

Strong bar magnet

Paper clips, steel wool, steel
  straight pins (not brass
  pins)

SMALL-GROUP ACTIVITY:

1. Let children use the magnet to determine if needles
   or clips are made of steel.
2. "Try to pick up steel wool bits with the needles or
   clips. Are they magnetic?" (They won't be yet.)
3. Show children how to pull the needle *across one end*
   of the magnet *in one direction.* Count aloud about 25
   strokes.
4. "Now try to attract steel wool with the needle. It
   should be magnetized. It has become a *temporary
   magnet.*"

Note: This is a good time to demonstrate how jarring a magnet weakens it. Hit the magnetized needle against something hard a few times. Now try to pick up some light steel object.
The pull will be very weak.

CONCEPT: **Magnets are strongest at each end.**

1. Which parts of a magnet are strongest?

LEARNING OBJECTIVE: To notice that the ends of a magnet act differently from the middle of the
                      magnet.

MATERIALS:

Bits of steel wool

Paper clips, steel key chain or
  light switch pull chain
  about 3″ (7.5 cm) long

Horseshoe and bar magnets

SMALL-GROUP ACTIVITY:

1. Hold the curved end of the horseshoe magnet over
   steel wool. Notice if any steel is attracted to the
   middle of this magnet. "Now hold the *ends* of the
   magnet over the steel wool. Do the *ends* act differently from the middle?" Try with a bar magnet.
2. Touch the chain with both ends of a bar magnet.
   Notice whether the whole chain clings to the mag-

GETTING READY:

Cut the steel wool into fine bits. (Did some of the steel cling to the scissor edges? The scissors became a temporary magnet.)

Open the key chain full length.

net or if the middle part dangles free. (Chain should be longer than the bar magnet.)

3. Dangle the key chain 1/4″ (1 cm) above the center of a bar magnet. (The pull of a strong magnet will visibly curve the end of the chain toward one end of the magnet.)

CONCEPT: Each end of a magnet acts differently.

1. Are magnets alike on both ends?

LEARNING OBJECTIVE: To notice that each end of a magnet acts differently, and that like ends of two magnets push away while unlike ends pull together.

MATERIALS:

2 strong bar magnets or 2 horseshoe magnets

String

Nail polish or tape

GETTING READY:

If using lightweight magnets, tape the string to a tabletop or chair seat.

Tie the string to any horizontal bar or chair back that will allow a heavy magnet to swing freely.

SMALL-GROUP ACTIVITY:

1. Tie the string to the center of a magnet, balancing it.
2. Let the magnet hang and swing freely. When it stops moving, one end of the bar (or one side of the horseshoe magnet) will be pointing north. Mark this end with tape or a dot of nail polish. Do the same for the second magnet. (This is why a magnetic compass works. Some children may know about compasses already. Others may find directional discussion confusing at this point.)
3. Let the children hold a magnet in each hand, then try to touch like ends (north to north, south to south). "What do you feel? Try to touch unlike ends. What happens?"

Give children lots of time to explore the attraction and repulsion effect. With strong magnets, results are fascinating. To retain strength, make a point of storing bar magnets in pairs with opposite poles touching.

Note: In this instance, using the correct scientific term can create more confusion than clarity. The ends of magnets are called the north (seeking) pole and south (seeking) pole. Adults find it confusing that the North *(geographic)* Pole is not in quite the same location as the Earth's north *(magnetic)* pole. Young children hold firmly to yet a different abstraction about the North Pole.

2. Does shape change magnetic force?

MATERIALS:

Play dough

Table knife

3 ring magnets

Disc magnets

LARGE-GROUP ACTIVITY:

1. Show ring, disc, and horseshoe magnets. Learn what children think about the positive and negative ends of these magnets. (It can be confusing to perceive the sides of a flat disc as the ends of a magnet.)

FIGURE 11–1

2. Point to each end of a horseshoe magnet. "This magnet was bent, bringing the two ends close together to make the pulling power stronger. Let's find out how a disc magnet is made, using a play-dough model." Roll a handful of dough into a long cylinder. "A long magnet is made first. The power is in each end. Where are the north and south ends?"

3. Cut the cylinder in half. "Now there are two magnets, each with north/south ends. The power collects at each end, no matter if the magnet is long or short." (See Figure 11–1.)

4. Cut a narrow slice from a cylinder. Find the north/south ends of the still vertical slice. "The power still collects in each end. Ring and disc magnets are slices of a longer magnet."

5. Stand several disc or ring magnets side by side, N/S poles together to form a cylinder. Then "float" three ring magnets by slipping them onto a vertically held pencil, with like ends facing to repel. "The power is always in the ends, no matter how they are turned." Let children explore an assortment of disc and ring magnets' power to attract paperclips and other magnets.

PRIMARY INTEGRATING ACTIVITIES

Math Experiences

Magnet in the Grab Bag. Half fill a large grocery bag with used steel bottle caps, common nails, paper clips, or other small steel objects. Use a string to tie a strong magnet to a stick, fishpole style. Let children take turns dipping the magnet into the bag, then counting aloud the number of objects they have pulled up. A numeral recognition version of this game can be played by cutting out colored paper fish, writing a numeral on each, and fastening a paper clip or safety pin to each fish.

Plastic numerals and counting shapes are made with magnets to use on coated steel bulletin boards. Use them to form sets of objects, labeled with corresponding numerals. Make a game of this by letting children draw the numerals from a paper bag, choose the corresponding quantity of objects, and place them on the bulletin board. You may have an old tin-coated steel baking sheet in your kitchen that would substitute well for the commercial steel bulletin board.

Music

Mention to children that we couldn't hear sounds from a stereo, VCR, radio, television set, or tape recorder without magnets. Recording tape is coated with magnetized powdered iron, and the other four devices have magnets in their mechanisms.

Sing the "Magnet Song."

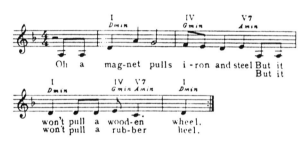

Magnet Song
(To the tune of: "The Cat Came Back")

—J. H.

Stories and Resources for Children

ARDLEY, NEIL. *The Science Book of Magnets.* San Diego: Harcourt Brace, 1991. The clear, well-organized text and clearly illustrated experiments make this book intriguing and useful for the independent reader.*

CHALLAND, HELEN. *Experiments with Magnets.* Chicago: Childrens Press, 1986. Carefully presented information about the makeup of magnets and naturally occurring magnets, as well as experiments to understand the force of magnetism.*

CHALLONER, JACK. *My First Batteries and Magnets Book.* New York: Dorling Kindersley, 1992. Clear instructions and excellent color photographs of magnet explorations and interesting toymaking have their own pulling power for children.

JENNINGS, TERRY. *Electricity and Magnetism.* New York: Smithmark, 1992. Familiar objects that depend on magnetic force to operate, and magnet projects and toys to make are featured in this book.

WHALLEY, MARGARET. *Experiment with Electricity and Magnetism.* Minneapolis: Lerner Publications, 1994. Standard beginning experiments are offered, plus a timely awareness of the potential harm magnets can cause to electronic equipment.

---

*Starred references, written at the young child's level of understanding, can help teachers who have minimal backgrounds in science to expand their knowledge base.

Storytelling with Magnets

Tell a story featuring practical uses of magnetic gadgets that could also be props for the telling. A small magnet could save the day when father drops his keys down the register, when sister's box of bobby pins spills in her bubble bath, when someone upsets a box of pins, or when a glass jar of nails breaks on the garage floor. These and other calamities could be resolved with an upholsterer's tack hammer, a magnetic-holder flashlight, a paper clip or pin box with a magnetized top, or other magnetic equipment that may be available.

The effects of magnetism can be used throughout the school year as a storytelling device in the following three ways:

Magnetically Directed Puppets. Tape a paperclip to the bottoms of small wooden dollhouse dolls. Put an open shoe box on its side to make a platform for the puppets (Figure 11–2). Make the puppets move by holding magnets in each hand beneath the platform. (Puppets can be thread spools with button heads glued on, clothespin dolls with paperclip feet, or pipe cleaner dolls.) Children love using magnet puppets to retell familiar stories or to create their own plays.

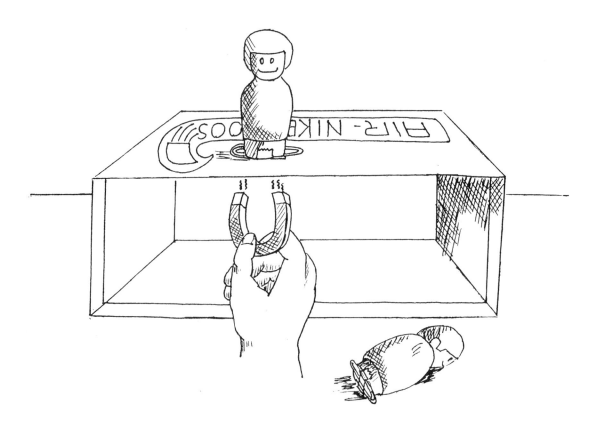

FIGURE 11–2

Paper Doll Stories. Cut out paper figures to illustrate a story. Tape paper clips or small safety pins to the backs. Invert a large grocery bag (plain or decorated) as a backdrop, and manipulate the paper figures with magnets held inside the bag.

Magnetic Bulletin Board Stories. Use fabric or paper figures as you would for a flannel board story. Hold them in place with a shirt button-sized magnet. Try to fit one of the whimsical magnet insects into the story.

Art Activities

Make junk sculpture with iron or steel.

1. Prepare steel bottle caps and cocoa tin lids by punching two nail holes into each one.
2. Prepare sculpture bases using mounds of damp clay, chunks of foam, or cardboard fruit trays.
3. Put out a tray full of iron and steel discards (paper clips, nails, steel bottle caps, hairpins, pipe cleaners, cocoa tin lids, twist bag closures, washers, strips of screening, soft wire). Put out several magnets so that the children can test materials for magnetic attraction. Explain that the sculpture for today will be made only of iron or steel.
4. Show how wire can be threaded through the punched bottle caps, how paper clips can be opened, and how wire pipe cleaners can be coiled around a finger to make interesting shapes.

Directions for magnetically manipulated painting with steel objects appear in *Mudpies to Magnets,* by Williams, Rockwell, and Sherwood.

Dramatic Play

Automobile Service. Tie a small magnet to a toy tow truck for children to use in hauling steel cars to the garage.

Magnet Table Games

Name Game. Print capital letters on separate small squares of paper. Put a staple through each square. Pile the squares into a small basket. Let the children fish for those letters contained in their names, using a magnet tied to a stick on a string.

Magnet Construction Game Box. Collect three or four small, strong magnets with keeper bars (or nails to serve that purpose). Find iron and steel discard items similar to those suggested for the junk sculpture. Try to include steel key chains, notebook rings, old keys, and cocoa box lids. Children enjoy combining odd shapes that are held together by magnetic attraction. Try to find a tin cookie box for storing the game. The lid and the box can serve as bases for the constructions, since the tin is a coating for steel.

Commercial magnetic building sets are available through school equipment catalogs. Magnetic games, magnet sculpture sets, and "Magnet Marbles" are available in toy and speciality stores. The Magna Doodle toy uses a magnetic wand to make iron filing designs on a screen.

Creative Thinking

I Am a Magnet. Collect a tray of assorted objects: key, steel bottle cap, spool, stick, rock, nail, bolt, paper clip. Sit with the children in a circle on the floor. Place the tray in the center. Start off as the leader when introducing the game. Ask the child next to you to choose one of the objects to pretend to be. Say, "I am a great, strong magnet. What are you, Maria?" If Maria replies that she is a bottle cap, say, "Then we'll cling together," and hold her hand. Encourage Maria to say to the child next to her, "Josh, I'm a magnetized bottle cap. What are you?" If Josh decides to be a rock, Maria goes on to the next child until she finds someone to cling to. End the game with, "Now I am a teacher again and my pulling strength is gone."

SECONDARY INTEGRATION

Maintaining Concepts

Point out magnets in use around the school whenever they come to your attention: refrigerator door sealing strips, car seat belt clasps, fancy buckles on children's belts, cupboard door latches. For the latter example, it is a good idea to suggest that children using those cupboards try to close them gently, without banging them, so that the magnets won't lose their strength. Show how the force of magnetism is used for banking, shopping, and other purposes by pointing out the magnetic strip on an ATM or credit card. Library, department store, and airport security check points are magnetic systems.

Make magnets a standard part of classroom cleanup equipment. Hang a magnet and keeper bar on a low peg near the workbench if you have one. Let the children use it on the floor, in workbench drawers, or on the workbench to gather up stray nails. Use it to sort out tiny nails from wooden shapes after children have worked with hammer-on design kits. Use the magnet to locate small steel cars that have become buried in the sandbox.

Connecting Concepts

1. When discussing the effect of magnetism passing through materials, ask the children whether air is one of the substances that magnetism passes through. How can they tell? Recall with the children that even though we cannot see air or magnetism, we have discovered that both are real things.

2. Set up a water play game that depends upon magnetism's passing through water. Let the children make barge-shaped boats from pressed foam, and fasten a paper clip to it with a rubber band. Stack three blocks under each end of a shal-

low cake pan. Fill the pan with water and launch the foam boats on it. Children can guide their boats through the water with magnets held beneath the pan.

3. Combine buoyancy and the principle of attraction and repulsion of like magnetic poles in a water play activity. Help the children magnetize 2" (5 cm) blunt needles, then place them on top of barge-shaped foam boats. Float the boats in a pan of water. Push the boats ahead by approaching the end of the needle with the like pole of a magnet; pull boats back with the unlike pole of the magnet. If a needle rolls off a boat, the children can go fishing for it in the water with a magnet.

RESOURCES

BERGER, MELVIN. *The Science of Music* (pp. 123–124). New York: Thomas Y. Crowell, 1988. Learn how microscopic magnets turn into audiotapes.

HANN, JUDITH. *How Science Works* (pp. 104–113). Pleasantville, NY: Reader's Digest, 1991.

LEVENSON, ELAINE. *Teaching Children About Physical Science* (pp. 71–90). New York: Tab Books, 1994.

VAN CLEAVE, JANICE. *Magnets*. New York: Wiley, 1993. Intended as a science fair projects book for older children, a good range of background information on magnetism is also offered.

WILLIAMS, ROBERT, ROCKWELL, ROBERT, & SHERWOOD, ELIZABETH. *Mudpies to Magnets*. Mt. Rainier, MD: Gryphon House, 1985. See "Magical Magnet Masterpieces" art project on page 87.

The Effects of Gravity

Children who have felt the tug of invisible magnetic attraction with their own hands can move from this awareness to simple understanding about the effects of gravity. These children are ready to put credence in the far stronger invisible pull that holds people, houses, and schools on the ground—the force of gravity. The learning experiences in this chapter amplify one central concept:

- Gravity pulls on everything.

Drawing children's attention to an ever-present effect that is rarely noticed or labeled calls for more preliminary description than we usually offer to action-loving children. The first suggested gravity experience is a story that provides basic information. The active experiences involve measuring and comparing gravity's pull on objects, trying a pendulum, and bringing things into balance.

Introduction: Lead into the gravity story by crushing a piece of paper into a ball and holding it between your fingers. "What will happen to this ball of paper if I turn my hand over and open my fingers?" Find out if the children's predictions are correct. "Do you think the ball might do the same thing another time, or might it fall up instead? Did you ever stumble and fall up? Here is a story about the reason for this."

Use a flannel board and felt figures to illustrate your story. It could be stated something like the following:

Once a girl had a dream. Everything seemed very strange. The girl was floating in the air looking for her house. She saw her friend floating close by with a ball in his arms. They wanted to play catch, but when the boy threw the ball, it drifted up out of reach. Then the girl saw her house bobbing up and down gently in the breeze. Her mom was very upset because somebody had spilled milk all over the ceiling. When the girl woke up, she was glad that her house was standing still.

Something important was missing in that dream. Houses and balls and children don't float around. Milk doesn't spill up! There is a reason why those things don't happen. There is a very powerful force pulling down on everything in the world: the force called Earth's *gravity*. We can't touch gravity or see it. We can just see what gravity does.

"Can you think of something else invisible that pulls some kinds of things? Gravity reminds us of the way magnets pull on iron and steel. But gravity is much, much stronger than magnetism. Gravity pulls from inside the Earth. *It pulls on everything all of the time.* We are so used to it that we don't even notice it happening."

(Show a picture of an astronaut in his space suit.) The astronauts who walked on the moon had a strange experience. They had to wear very heavy boots to stand on the moon, because the moon's gravity pull is weaker than the Earth's gravity pull. At the science table you can find out how much Earth's gravity is pulling on you.

CONCEPT: **Gravity pulls on everything.**

1. Can we experience gravity pulling on us?

Invite a volunteer child to be the demonstration subject to introduce this activity. Have the child sit on a straight chair. Make sure both his feet are flat on the floor, his back is touching the chair back, and his hands are in his lap. Now ask the child to try to stand up without swaying his body forward, nor moving any other muscles. . . not even a tiny bit. Encourage the child to try very hard to stand up, then report to the class what he is experiencing. Ask, "What will you have to do in order to stand up? Think about it."

Let the rest of the class take turns trying the experience in their small groups, while others observe for motionlessness. Afterward discuss what the children noticed. "What do you think held you to the chair when you didn't move a muscle? What did you need to do to stand up?"

Feeling gravity's inexorable pull.

Guide the discussion to the conclusion that it is gravity's force that holds us down in the chair. The force of gravity always pulls on everything. We have to use our energy to push ourselves out of the chair and stand up. It takes energy to move the muscles we use to stand up. It always takes energy to work against gravity's pull.

2. How much does gravity pull on different objects?

LEARNING OBJECTIVE: To experience the way gravity pulls on one's body.

MATERIALS:

Bathroom scale

Kitchen scale

4 small, sturdy bags

Sand, pebbles

Market bag with handles

Containers, scoops

Long mirror, if possible

GETTING READY:

Fill each small bag with 2
  pounds of sand or stones.

SMALL-GROUP ACTIVITY:

1. "When we weigh ourselves on a scale, we are find- ing out how much gravity pulls on us."
2. Find out how much gravity pulls on each child. Record the weights.
3. Weigh each child again while she holds a sand bag in each hand. Compare with first weight record. Which way does gravity pull more?
4. Put both sand bags in the market bag. Have chil- dren hold it in one hand so the weight is on one side. Weigh each child again.
5. "Look in the mirror. Are you standing up straight now or leaning over? You lean away from the heavy side to keep your balance."

Let children weigh objects on both scales. Containers of sand make good materials to weigh. Provide objects of varying weight and size. Let children discover that size does not always determine weight.

Group Discussion: Recall the earlier discussion about gravity holding everything down. Encourage children to share ideas about things that they see being pulled down from the sky by gravity (leaves, seeds, snowflakes, raindrops).

Mention that some things go up and move through the air for a long time (planes, birds, gliders, and so on). Energy from wind, motors, and jet engines can lift planes up and keep them up for a while. Can the children lift themselves all the way off the ground for a little while by using their energy to push against gravity?

3. Can we compare gravity's pull on objects?

LEARNING OBJECTIVE: To explore ways of achieving equilibrium by using varied small objects as weights in balance cups.

MATERIALS:

Scrap lumber: 2' (60 cm) piece
  of dowel or broomstick; 8"
  (20 cm) × 8" (20 cm) piece of
  shelving

Nails: 6d and 4d sizes

Yardstick (meterstick)

Pipe cleaners

2 small, reused yogurt cups

SMALL-GROUP ACTIVITY:

1. To make base: Mark center of shelving square. Nail dowel perpendicular to center of base. (Let chil- dren do the hammering.)
2. Nail yardstick (meterstick) loosely near the top of the upright stick, driving a 4d nail through hole at the 18" (50 cm) mark. Yardstick (meterstick) should swing easily.
3. Insert pipe cleaners into yardstick (meterstick) holes. Hook them into paper cups. (Put one cup on

Chris uses the yardstick balance she helped to assemble.

Small objects to weight: shells, washers, acorns, etc.

GETTING READY:

Drill holes or saw small notches 3" apart on yardstick (8 cm on meterstick).

Make a hole at 18" (50 cm) and mark with the 6d nail.

Make 2 holes on opposite sides of each container

each side. Let children add more cups as they wish.) If notches were made instead of drilled holes, make pipe cleaner handles for cups and hang in notches.

4. Let children fill cups with small items to balance.

Provide materials as different in weight as cotton balls and shells. Do five cotton balls balance five shells? Is the balance arm lower on the shell side? What does this mean? (Gravity pulls more on shells, thus we say they weigh more than the cotton balls.)

Try to keep the balances and containers of weighing materials accessible to the children for as long as they show interest in them. Put them away for a while; bring out periodically.

Group Experience: Tie a small doll at the end of a foot of string. Hold the other end of the string, letting the doll dangle. "How could I make the doll go higher without lifting up the string? Good, I could swing the doll higher. Look, it is going high now, but it doesn't stay high. Why? Let's watch what happens when I stop swinging the string. Is it moving up as high now? It's going slower and lower, and it has almost stopped swinging. What could be helping to stop it?

"Is this what happens to you when you're on the swing? What happens to you when you start to pull and push with your muscles? What happens when you stop using your energy to keep the swing moving back and forth? Are you close to the top of the swing set or close to the ground when the swing stops? Why?" (The force of gravity is part of the reason.) Tape the string and doll pendulum to the edge of a table for children to try later.

Continue the pendulum investigation another day, making a sand pendulum so that children can observe the pattern of the pendulum arcs. Make a funnel by cutting off the bottom of an empty dish detergent bottle. Make three evenly spaced holes near the cut edge. Tie three strong strings to the holes; bundle the strings and tie them to a sturdy cord (Figure 12–1). Suspend the upended funnel by the cord, from a doorframe or ceiling beam. The capped end of the funnel needs to be close to the floor, about an inch above, so that the sand doesn't bounce.

Spread a 3′ x 2′ sheet of paper or plastic under the pendulum. Half-fill the funnel with sand. Pull the cap open, give the funnel a push, and watch the sand flow into interesting arc patterns below. "Do the swings start to go out wider and wider, or do they get smaller?"

To make the sand flow more smoothly, remove the bit of plastic that has been put there to slow down the flow of detergent. (See Math Experiences to ex- this activity.)

4. Can we keep objects in balance?

LEARNING OBJECTIVE: To explore ways to achieve equilibrium by placing rocks on a ruler that is balanced on a block.

MATERIALS:

12″ ruler

Half-circle blocks

Small rocks

Tinkertoys

Desirable: Acrobat balance toy
   sold by import gift shops

SMALL-GROUP ACTIVITY:

1. Gently place a ruler on a block, with the 6″ mark resting on top of the curve. "Let's see if the ruler will tip or balance. Can you try this?"
2. "Can we balance two rocks on the ruler?" Show how to slide a heavier rock toward the center (resting point) to balance a lighter rock at the other end of the ruler.
3. Make a platform of two Tinkertoy wheels and one long stick. Insert two sticks in a connector wheel and add small connectors to the stick ends. Rest connector wheel on the platform edge as shown in Figure 12–2.
4. Remove one stick, then the other, from the connector wheel. What happens? Is the connector wheel stable without them? (The location and weight of the sticks change the way gravity pulls on the connector wheel.)

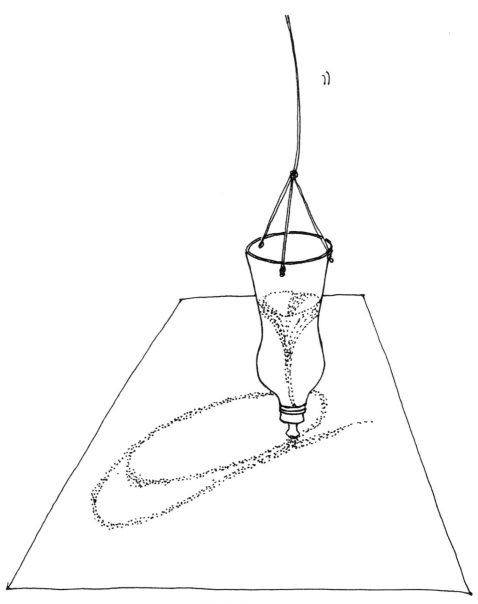

FIGURE 12–1

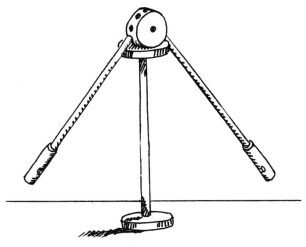

FIGURE 12–2

> **REMEMBER**
>
> Respect the questions children ask.
> Let your questions encourage finding out.

PRIMARY INTEGRATING ACTIVITIES

Math Experiences

When they experiment with a pan balance and weights, children can strengthen their understanding of numerical equivalence. Sets of standard weights are available commercially. In some types, the size and weight are related. To devise substitute weights, fill screw-top plastic or metal containers with sand and gravel to equal 1-, 2-, and 4-ounce (25-, 50-, and 100-gram) weights as desired. Mark the corresponding numeral on each container.

Number balance equipment is made in several forms. One type uses weighted plastic numerals to make the mathematical equation balance as the weight balances. Keep a set of weights, along with boxes of materials to weigh, next to a pan balance for children to use when time permits.

Older children can count the number of pendulum swings per minute. Compare the rate of swings per minute for a pendulum on a short string, and one on a long string.

Michelle compares gravity's pull on objects.

Music

Listen to Michael Mish's "Gravity" song on his *I'm Blue* cassette.

When children are investigating the effects of gravity, they enjoy singing three nursery rhymes as though gravity were not operating. "Jack and Jill Went Up the Hill" might be sung ". . . Jack fell *up,* and broke his cup, And Jill bent over with laughter." "London Bridge" might be sung ". . . London Bridge is *floating up,*" and "Ring Around the Rosy" might be sung ". . . One, two, three and we fall *up* in the tree."

Here is a gravity song to sing.

Gravity Song

What keeps peo-ple on the ground, what keeps us from

float-ing up? What keeps ev'-ry-thing down? Grav-i-ty, that's what!

—J. H.

Stories and Resources for Children

BOONE, EMILIE. *Belinda's Balloon.* New York: Alfred A. Knopf, 1985. Belinda's balloon lifts her off the ground. Her sister helps solve the problem of becoming heavy enough to drift back to Earth.

BRANLEY, FRANKLYN. *Gravity Is a Mystery.* New York: Harper & Row, 1986. This book helps children learn what gravity does. It doesn't explain just what gravity is, because no one has been able to explain it yet. Paperback.*

COBB, VICKI. *Why Doesn't the Earth Fall Up?* New York: E. P. Dutton, 1988. Gravity and the laws of motion are explained simply.*

CONAWAY, JUDITH. *More Science Secrets.* Mahwah, NJ: Troll, 1987. Includes directions for making a "whirling gravity ball" to illustrate how the sun's gravity holds Earth in its regular orbit. Paperback.

EVANS, DAVID, & WILLIAMS, C. *Make It Balance.* New York: Dorling Kindersley, 1992. Vivid photographs invite the youngest explorers to try open-ended balancing challenges. Explanations of gravity's role are listed separately for adult guidance.

KNAPP, BRIAN. *Falling.* Danbury, CT: Grolier, 1991. Good photographs illustrate intriguing explorations of the force of gravity, and the ways we work with it and against it.

McCULLY, EMILY. *Mirette on the High Wire.* New York: G. P. Putnam's Sons, 1992. A venturesome girl persists and learns to concentrate and balance as a highwire walker. She saves a performer. A Caldecott Award winner.

McNULTY, FAITH. *How to Dig a Hole to the Other Side of the World.* New York: Harper & Row, 1990. Weightlessness is experienced at the center of the Earth in this fantasy journey.

MINARIK, ELSE. *Little Bear.* New York: Harper & Row, 1961. Chapter 3, "Little Bear Goes to the Moon." Paperback.

MOCHE, DINA. *If You Were an Astronaut.* New York: Golden Books, 1985. Daily life aboard a space shuttle is portrayed. Teachers need to specify that "Zero G" means no gravity pull. Paperback.

RIDE, SALLY, & OKIE, SUSAN. *To Space and Back.* New York: Lothrop, 1986. Young children will enjoy hearing the portions of this book portraying the peculiarities of doing ordinary life routines weightlessly in an orbiting space capsule. Excellent color photographs of the astronaut/author in space.

ZUBROWSKI, BERNIE. *Mobiles: Building and Experimenting with Balancing Toys.* A Boston Children's Museum Book. New York: Beech Tree, 1993. Interesting experiences with body balance, and finding balance points with various objects, are included with directions for simple and complex balance toy making. Paperback.

Swinging, sliding, blockbuilding, and falling down are part of many other children's stories as well. Casually refer to the effects of gravity's pull whenever possible.

*Starred references, written at the young child's level of understanding, can help teachers who have minimal backgrounds in science to expand their knowledge base.

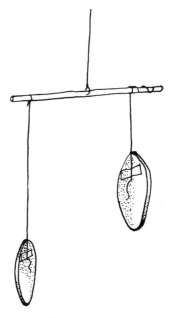

FIGURE 12–3

Poems (see Appendix A)

Gravity doesn't seem to operate in "The Folks Who Live in Backward Town," by Mary Ann Hoberman (in *Poems Children Will Sit Still For,* de Regniers, Moore, and White). Also read the classic poem about swinging, "The Swing," by Robert Louis Stevenson in *A Child's Garden of Verses*.

Art Activities

Create nature mobiles. Collect fallen sticks outdoors. Let children help, if the schoolgrounds have trees. Using tape and string, suspend each stick at a level that allows children to work on it. Provide two small objects from nature, such as cones and pods, to hang by a string from each stick. Children can experiment to achieve a balance by moving the objects' strings closer or farther away from the center of the stick, or by winding one string around the stick to make it shorter than the other (Figure 12–3). Shapes cut from pressed foam plastic can also be used.

Dramatic Play

Store. Put a pan balance or old kitchen scales, containers of materials like horse chestnuts, and artificial fruit and vegetables in the play store area.

Block Play. Help block builders think about why their structures collapse. Suggest making broad, sturdy bases for tall towers, so that gravity will pull more on the bottom of the tower than on the top. When appropriate, ask if gravity is pulling equally on both sides of the weight-supporting blocks or whether an unbalanced load will tip over.

Spaceship. Try to find a refrigerator shipping carton for the children to convert into a spaceship. Help them improvise space helmets by using small cartons and coils of telephone cable wire, like the one worn by Little Bear on his imaginary trip to the moon. Before setting up the spaceship play, supply play ideas by reading *Little Bear* by Else Minarik or another one of the astronaut books.

Creative Movement

Help the children recognize, when they dance or exercise, the way their bodies involuntarily adjust to changes in body position to maintain their balance. When they lean in one direction, their bodies automatically compensate for gravity's pull by extending an arm or leg in the opposite direction.

Let the children pretend that they are on a ship in a storm, rocking from side to side. They will stretch their legs into a wide-based stance or tip over. Suggest moving like ice skaters, swinging arms and bending bodies as they glide and stride on the ice. Now they are tightrope walkers sliding one foot in front of the other on a swaying rope, arms extended. Next, imagine that they are going to lift a heavy rock, working very hard with their muscles to force it up against gravity's pull, bit by bit. Now put the rock down carefully and push it ahead slowly. Then move like the astronauts on the moon where the gravity pull is weak.

In *Creative Movement for the Developing Child*, Clare Cherry suggests two poems as stimuli for balancing movements: "The Cat on the Fence" and "The High Wire Walker."

Whenever tension seems to be mounting in the classroom, or when excited children need help shifting from physical activity to a learning situation that requires concentration, use this "anti-gravity relaxation fantasy." Guide it slowly and quietly, saying: "Let your eyelids slide down now to close . . . and let your hands be loose and comfortable in your lap . . . Let your arms be loose . . . and limp . . . and light . . . so light they seem to float . . . so light that it seems that gravity isn't pulling on them much at all . . . And now . . . notice . . . just notice how loose and light your neck and shoulders feel . . . There is just enough gravity pull to help them keep your head above them . . . But your head might want to move alllll around slowly, till it finds a comfortable place to rest. There, it feels soooo comfortable now . . . There is almost no heaviness in your body now . . . Everything feels easy now . . . Let your mind take you on a floating journey now . . . You might want to have a cloud to rest on, feeling all that softness around you . . . or you might want to be an astronaut floating inside a space capsule, far, far, away from Earth's gravity pull . . . It's your floating journey, so you can have it be just the way you want it to be, because gravity doesn't pull on your thoughts—ever . . . so you can float as easily as you like in your imagination . . . And then gradually, slowly, you begin to notice gravity's pull again . . . You notice how well you are breathing now as you drift down closer, closer to Earth . . . Feeling rested and fresh and relaxed now . . . And you take three long, slow, deep breaths . . . and then your eyes open again, so that you can see what we will be doing next in school."

Creative Thinking

Use gravity pull as a topic to stimulate creative thinking. Ask, "What if there were no gravity pull from Earth? What would be different? What would it be like to play on the playground? What if we had to eat our meals as the astronauts do on space-flights? Would we be sitting in our room?"

Food Experiences

Take the class to the nearest food store where purchases are weighed on a balance beam scale. Buying and sharing a half pound of peanuts with the children is a good investment in gravity learning.

Field Trips

Knowing how much gravity pulls on objects is an important part of many businesses and services. Try to include a look at weighing devices during the field trips you plan to the post office, grocery store, feed store, airport, drugstore, medical offices, or the loading docks of factories. It can be great fun for a whole class of children to be weighed together on loading dock scales.

SECONDARY INTEGRATION

Maintaining Concepts

One child applied gravity concepts in a judgmental way when he picked himself up from a fall complaining, "That ol' grabbity pulled me down!" Many of the large muscle-play activities provide opportunities to point out how gravity's pull makes some of our fun or work easier, such as balancing block towers, enjoying a slide, teeter-totter, or swing, playing catch, or pouring water from a pitcher. Some activities are difficult to do because of gravity's pull. This is why children become tired when they put away blocks, climb the jungle gym, walk up the stairs, or trudge up a hill.

Connecting Concepts

Gravity/Air Relationship. Balancing two air-filled balloons is a vivid way to demonstrate that gravity pulls on air. To make a long balance arm, insert one drinking straw partway into another straw. Insert a hanging loop in the center as you did in making the mobile. Blow up two balloons to equal size; tie one to each end of the arm. Be sure that the balloons are the same color. Some children may believe that balloons of different colors are also different in the amount of air they contain.

Hang up the arrangement and let the arm come to rest. Add bits of masking tape to the end of the arm that is higher until balance is achieved. It is hard to blow exactly the same amount of air into two balloons. Explain to the children that you are going to prick one balloon. "What will happen to that balloon?" Some children become scared when the balloon pops and forget to watch what happens to the balance. "Put your hands over your ears if the noise might bother you." Pop the balloon. "Which side of the balance went down? Why did that happen? Gravity pulls on air, too!" (If the pricked balloon flies apart, gather the pieces and tape them onto the end of the balance stick so a comparison of empty and air-filled balloons may still be made.)

Gravity/Water Relationship. Introduce the idea that gravity's pull plays a part in determining which objects sink and which float. Gravity pulls on water more than it pulls on corks, for instance. Recall that raindrops are pulled down to Earth when they become too heavy to float as droplets of vapor in clouds (see Chapter 9, Weather).

Family Involvement

Families often have opportunities to show their children things that go up and away from gravity's pull. Big motors help elevators and escalators lift people. Jet propulsion or propellers help lift airplanes off the ground. Heavy motors are needed to lift loads of materials at building construction sites that families could visit together.

RESOURCES

HANN, JUDITH. *How Science Works.* Pleasantville, NY: Readers' Digest Association, 1991.
LEVENSON, ELAINE. *Teaching Children About Physical Science.* New York: TAB Books, 1994.
WILLIAMS, R., ROCKWELL, R., & SHERWOOD, E. *Mudpies to Magnets.* Mt. Rainier, MD: Gryphon House, 1987. Includes directions for making bleach bottle "space helmets" and for making a balance toy.

Simple Machines

Throbbing motors and turning wheels make little-noted daily background noises in the Western world. Children are proud to learn how to move and lift things with the help of simple machines. The knowledge invites interest in the complex machines they have previously taken for granted. The learning experiences in this chapter explore the following concepts:

- Friction can heat, slow, and wear away objects.
- A lever helps lift objects.
- A ramp shares the work of lifting.
- A screw is a curved ramp.
- Simple machines help move things along.
- Some wheels turn alone; some turn together.
- Single wheels can turn other wheels.
- Single wheels can help us pull down to lift up.

Experiencing the advantages and disadvantages of friction is a useful preliminary to the experiments that follow. The simple machines experiments illustrate the lever, ramp, screw, wheel and axle, and pulley.

Introducing Friction: "Rub the palms of your hands together as fast as you can. Keep going. Do you feel something happening to your hands? What we are doing makes heat. The reason for the heat is a force called *friction*."

CONCEPT: Friction can heat, slow, and wear away objects.

1. Does friction slow sliding objects?

LEARNING OBJECTIVE: To experience firsthand the slowing effect of friction on a slide.

MATERIALS:

Gum erasers

Small pieces of waxed paper

Magnifying glass

Indoor slide or teeter-totter
  plank, propped against a
  table

Sheets of waxed paper

Rubber sink or tub mat

LARGE-GROUP ACTIVITY:

1. Pass around pieces of waxed paper. "Try rubbing
   this on your arm. Does it slide fast? Try sliding the
   rubber eraser on your arm? Does it slide fast? Can
   you think of a reason for this?" (The rubber makes
   more friction.)
2. "Look at the paper and the eraser through the
   magnifying glass. Which is smooth? Which is
   fuzzy?"
3. Let children take turns on the slide, sitting first on
   waxed paper, then on the rubber mat. "Smooth paper doesn't make much friction, so you move
   fast."

2. Does friction make heat and wear objects away?

LEARNING OBJECTIVE: To experience firsthand the heat produced by friction and to actively wear
                     away materials by producing friction.

MATERIALS:

6" (15 cm) pieces of scrap
  lumber

Coarse sandpaper, cut into 3"
  (7 cm) squares

Hammer

Nails

Magnifying glass

GETTING READY:

Hammer a few nails half their
  length into a chunk of wood.
  (To use the hammer as a
  lever, catch nail in claw and
  roll hammer back on curve
  of claw. Pull *down* on handle
  to lift nail *up* and out.)

SMALL-GROUP ACTIVITY:

1. Give each child sandpaper and scrap lumber. "Feel
   these. Are they smooth or rough? Do they look
   smooth or bumpy with the magnifier? Rub them
   together hard and long. See what happens."
2. "Are your fingers feeling warm? (Rough things
   rubbing together make lots of friction.) Are bits of
   wood wearing away? Look at your wood. Feel it.
   Is it getting smoother? Is the sandpaper changing?
   What does friction seem to do to things?"
3. "I'm going to pull a nail from this wood. Do you
   think that will make friction? Let's feel the nail
   quickly after it comes out. Is it warm or cool? Do
   you think the nail and wood rubbed together?
   How do you know?"

3. Can we cut down friction?

LEARNING OBJECTIVE: To notice how lubricating materials reduces friction.

MATERIALS:

Coarse sandpaper

Petroleum jelly

SMALL-GROUP ACTIVITY:

1. Give children two pieces of sandpaper. "Rub these
   together fast. What happens to the bits of sand?"

Meat trays

Peanut butter

Soda crackers

Waxed paper

Table knives

GETTING READY:

Cut sandpaper into 2″ (5 cm)
   squares.

Put a spoonful of peanut butter on pieces of waxed paper
   for each child.

2. "Look at the sandpaper. Is it bumpy and rough?
   Do you think it would be smoother if something
   filled up the bumpy places? Try this petroleum
   jelly on it. See if the pieces of sandpaper slide past
   each other smoothly now."
3. Give each child two crackers on a tray. Repeat
   steps 1 and 2, offering peanut butter as a lubricant.
   Let the children eat this experiment.

Group Discussion: Show the children a small can of household oil. Ask what they know about
it and why it is used at home. Develop the idea that oil makes a smooth sliding surface on objects that rub together. It cuts down friction that makes heat, wears bits away, makes things difficult to use, and makes things squeak.

4. Can friction be useful?

LEARNING OBJECTIVE: *To notice that friction is needed for writing, drawing, and gripping things.*

MATERIALS:

Chalk

Dark-colored paper

Small pieces of waxed paper

Pencils

Petroleum jelly

SMALL-GROUP ACTIVITY:

A

1. Give children paper and chalk. "What sound does
   the chalk make on the paper? Scraping? Scratching?
   Why do you think this happens?"
2. "Try dipping your chalk into the petroleum jelly.
   Now does it rub and scrape? What happened to the
   chalk mark? Do you need friction to draw?"
3. Let them try to draw on waxed paper with pencils.
   Explain that the wax coating is like a lubricant.

B

Screw-top plastic containers,
   such as small paste jars

Two pans of water

Soap

Towels

1. "Are your hands strong enough to unscrew this jar
   cover? You can do it well."
2. Try it again, but this time get your hands wet and
   soapy first. Why is it hard to do now? What is missing?"
3. "Try the doorknob with wet hands. Can you turn
   it?"

Group Discussion: Ask the children to check each other's shoe soles. Are some smooth and
slippery? Do some have rubber heels? Do others have one-piece rubber bottoms? Are sneakers good for running and stopping, or do they skid? Are car tires smooth and slippery, or
more like the soles of sneakers? Why? Talk about slippery road conditions, slippery bath-
tubs, and wet bathroom floors. How does friction make these places safer?

CONCEPT: A lever helps lift objects.

1. What can a lever do?

LEARNING OBJECTIVE: To discover how to arrange a plank and resting point as a lever to lift and be lifted.

Introduction: "One day I watched Sophia on the playground pushing *down* to lift David *up* in the air. Then David pushed *down* to lift Sophia *up*. What were they doing? (Using a seesaw.) Could Amanda lift me that way? Gravity pulls much more on me than on her. Let's see if we can find a way."

MATERIALS:

Six-foot seesaw plank

Block, books

12" rulers or metersticks

Pencils

String

Hammer, nails

Scrap lumber

Sturdy box large enough to
hold two seated children

SMALL-GROUP ACTIVITY:

1. Put a block below midpoint of the plank. Have a child stand on one end. Ask another to stand on the raised end to lift the other.
2. "Could one of you lift me up that way? Try. Now let's see what will happen if we move the resting-point block closer to me. Now can you lift me? Yes! We changed the resting point and made a lever. It helped her lift me *up* when she pushed her weight *down*."
3. To show the importance of the resting point, place the box on one end of the plank. Put the block nearby. Ask two children to sit in the box. "How could we use the plank to lift this load?" The load tips if the *plank is lifted up.* The load is *lifted up* as the plank is *pushed down* only when the plank and block are used together as a lever and a resting point.

Children who are waiting for turns to lift the box with the plank lever can try out two other levers. Tie two or more books together with a string. Loop one end of the string so children can try to lift the bundle with one finger. Offer a ruler and a pencil for a resting point to make a lever. "It was hard to lift up the bundle pulling *up* with one finger. Try pushing *down* on the ruler lever to lift the books up."

Drive nails into the wood about half the nail length. "It can be very hard work to pull a nail out by pulling up. Try using the hammer as a lever, pushing down to lift the nail up."

Group Discussion: In *Simple Machines* by Rae Bains, look at the picture of a cave dweller using a branch and a small rock to move a huge rock. Pry open an empty cocoa can with the handle of a spoon. Ask if the spoon was a lever. Show that the resting point was the edge of the can. Talk about oars on a rowboat serving as levers that are pulled *back* to push the boat *forward.* Sing "Row, Row, Row Your Boat."

CONCEPT: A ramp shares the work of lifting.

The inclined plane (ramp) is another simple machine that helps us lift things against gravity's pull.

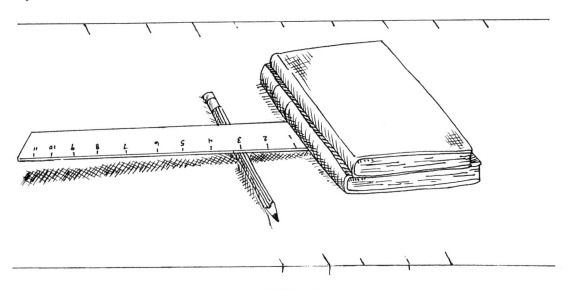

FIGURE 13–1

1. Can a ramp help us lift?

LEARNING OBJECTIVE: To find out firsthand if a ramp helps lift a heavy load.

MATERIALS:

Old, sturdy suitcase

Heavy objects to fill suitcase

Plank, at least 4′ (1.2 m) long

Table

GETTING READY:

Fill the suitcase with enough
  weight to make it hard for a
  child to lift by the handle.

SMALL-GROUP ACTIVITY:

1. "See if you can pick up this heavy case by the handle and lift it onto the table."
2. "See if this plank can make it easier to get the case up on the table." Lean the plank from the floor on the tabletop. Place the suitcase near the bottom on the plank. "We call this a ramp."
3. "A ramp holds some of the weight of the suitcase as you slide it along. Now you can get the suitcase up there."

Group Experience: Make a miniature steep mountain of damp sand in a long cake pan or in the sand table if you have one. Use a tiny matchbox car as a demonstrator. "Let's see if a car can go straight up the side of a steep mountain like this. What will happen if I let go? Gravity pulls it down *fast*. Could we curve a ramp road around this mountain to help the car go up?" Make a road like this, then let children try the car on the road. What happens when the children let go of it? (It may stand still in the damp sand, but on a real mountain road, gravity could slowly pull a car down.)

CONCEPT: A screw is a curved ramp.

Introduction: Draw a thick crayon line diagonally from one corner of a sheet of typing paper to the other. Cut along the line, leaving a crayon-edged triangle. Bring the triangle, a pencil, a large screw, and a collection of nuts and bolts to a class discussion. "Does the line on this paper triangle look like a ramp? Let's see what we can do with it. " Wind the triangle around the pencil so the line looks like a screw thread. Trace the spiral path from the bottom of the ramp to the top. Compare it with a screw. Give children the bolts and nuts. "Can you make the nut go up the ramp that curves around the bolt? Try it."

1. Can screws lift objects?

LEARNING OBJECTIVE: To investigate how threaded objects move up the corresponding threads of a screw.

MATERIALS:

(Use as many as possible)

Screw-type cookie press

Tissue

C-clamps

Modeling clay

Old lipstick tube

Screw-type nut cracker

Screw-top plastic jars

Plumbing pipes and joints
  (now available as plastic
  toys)

Old piano stool or office
  swivel chair

GETTING READY:

The bottom half of a lipstick
  tube is usually two pieces.
  Separate them to reveal the
  spiral ramp that the lipstick
  moves along as it is twisted
  up. Show children how this
  device works.

SMALL-GROUP ACTIVITY:

1. Remove design plate from the cookie press. Stuff a tissue in the press. Let children see what happens when the screw knob is turned. "What lifted up the tissue?"
2. Put a small ball of clay on the end of the C-clamp, turning the screw. Hold clamp upright; turn the screw until the clay rides up to be squashed by the top.
3. Show how two screws can be threaded together so they can't be pulled apart (use jars and pipes).
4. Let children enjoy working with the small materials you have assembled before presenting the experience of being lifted by a screw on a piano stool or office chair. Take time to examine the emerging screw as it turns.

CONCEPT: Simple machines help move things along.

1. How do rollers move things?

LEARNING OBJECTIVE: To experience and compare the difference between dragging a box, pushing it over rollers, and pushing it with wheels under it.

MATERIALS:

Grocery cartons

Jump ropes

4 cut-off broomsticks or cardboard cores from newsprint
rolls to use as rollers

Platform dolly or board and 4
caster wheels

GETTING READY:

If the school custodian doesn't
have a platform dolly for scrub
buckets, make your own by
screwing four swivel casters
into a 1″ (2.5 cm) thick plank.

SMALL-GROUP ACTIVITY:

1. Loop jump ropes around cartons. Let children take
turns pulling each other in boxes, using jump rope
handles in each hand. (Cardboard boxes fall apart
soon, so have many.) "Is pulling easy or hard?"
2. Place rollers side by side on the floor. Put box on
top. Repeat step 1. "See if rollers help move the
box." (They don't stay under the box. Put them
back for the next child's turn.)
3. "Which way made less friction?" Feel box bottoms
after the dragging and after rolling.
4. "Now let's try rolling wheels that will stay under
the box." Put box on the platform dolly. "Which is
the easiest way to pull the box: dragging, with
rollers, or with wheels?"

Group Discussion: Recall the rollers experiment. "Was the carton moved very far by four
rollers? Would rollers be a good way to move cars and trucks? What if someone had to keep
putting rollers in front of the car to keep it moving? Rollers are used that way when whole
houses are moved a short distance. They give good support to the house. Rollers are also used
to unload big trucks at the grocery store. Rollers are part of conveyor belts and escalators."

CONCEPT: Some wheels turn alone; some turn together.

1. How do single wheels and pairs of wheels work?

LEARNING OBJECTIVE: To notice the difference between things that move on single wheels and
those that move on pairs of wheels joined by axles.

MATERIALS:

A light piece of furniture on 4
swivel wheels (office chair,
typewriter stand, utility
table, crate on casters, platform dolly)

Toy car with axles exposed

Small paper plates

Masking tape

Plastic drinking straws

Compass

GETTING READY:

Find the exact centers of paper plates with a compass.
Mark each center; with pencil tip, push out a 1/4″ (1
cm) hole. Have one for each
child.

SMALL-GROUP ACTIVITY:

1. Divide children into two facing rows, at least 5 feet
apart.
2. "Let's take turns pushing the chair to children
across the room. Watch what happens. How does
it move?"
3. "Now let's push these cars across. Do they move
like the chair wheels?" Compare with the swivel
again, watching the wheels very closely. Turn the
swivel and the cars upside down to see if they look
the same. Supply the word *axle* if a child doesn't
offer it. Point out pairs of wheels that turn together
on an axle.
4. Let children try to roll their plates across to another child. Offer straws and tape, suggesting that
two children could join wheel plates to see if they
will roll better as a pair with an axle.

Cut 1" (2.5 cm) pieces of tape;
    stick lightly on a nearby
    table edge for children's use.

Group Discussion: Talk about the advantages of single wheels and pairs of wheels. If your classroom has an old upright piano or a classroom storage chest on casters, have children push it. Tiny, single wheels can make it easier to push something very heavy for a short way.

"Do you think wagons, cars, buses, and trucks have a single wheel on each corner?" Show some pictures of wheeled objects cut from catalogs. "Do these things use single wheels or pairs of wheels on axles?" Children are surprised to learn that doorknobs are pairs of wheels on an axle. Bring in an old doorknob or unscrew one at school. Look together at a manual typewriter for rollers, wheels, and axles. Try to bring in a conventional roller skate to compare with the single wheels of in-line skates.

CONCEPT: **Single wheels can turn other wheels.**

1. How do gears work?

LEARNING OBJECTIVE: To observe how sets of gears mesh to operate mechanisms.

MATERIALS:

Eggbeater (hand operated)

2 deep mixing bowls

Soupspoons

Water

Detergent

Gear toys, visible clockwork,
    visible music box works, old
    clock to open

Crayon

Newspapers

Sponge

GETTING READY:

Put 1/2 cup water and 1 tea-
    spoon detergent in each
    bowl.

Spread papers on the floor.

SMALL-GROUP ACTIVITY:

1. "Let's look very closely at a beater. How many wheels do you see? (Some wheels may look different.) Watch what happens to the little wheel when the big wheel is turned." Introduce the term *gear,* explaining that this is a wheel with teeth that mesh with teeth on the wheels next to it.
2. "Let's take turns using the beater to find out how the gears work. Let's compare beating bowls of soap suds with the beater and with a spoon to find out which works more easily."
3. Help children see that the big gear turns the little gears on the beater blades much faster.
4. Make other gear toys available for independent inspection.

Note: Children enjoy using a screwdriver to take apart an old windup clock. However, the coiled steel spring is *sharp.* It should be unscrewed and removed only by an adult.

Group Discussion: Try to borrow a one-speed bicycle to examine with the group. Turn the bicycle upside down and look at the gears. "Sometimes gears turn each other without touching each other. Instead they fit into a chain that moves them. Is one gear large and one small, like the eggbeater? Let's turn the pedals and watch the chain make the small gear

Some wheels turn alone.

turn. Count how many times the back wheel turns while I turn the pedal one time." Count one turn when the air valve reaches the top.

Look at the tire tread. "Could friction help stop a bicycle? Why do you think the chain looks greasy? Would it need grease? What do you think is inside the tires?"

CONCEPT: Single wheels can help us pull down to lift up.

1. Does a pulley make lifting easier?

LEARNING OBJECTIVE: To experience by comparison how a pulley makes lifting easier.

MATERIALS:

Small pulley

Firm cord, double the length
  of floor-to-pulley distance

Small school chair

Stopwatch

Screw hook

GETTING READY:

Install screw hook in a ceiling
  beam or door frame (ask
  custodian).

SMALL-GROUP ACTIVITY:

1. Show children the pulley. "Do you suppose this
   single wheel could make it easier to lift a chair?"
2. "First let's see how long you can hold this chair with
   one hand. Lift it as high as you can. I'll time you."
   Record length of time each child can hold chair.
3. Tie cord firmly to chair back. "Now try pulling
   down on this cord to lift the chair. I'll time you to
   see how long you can hold it up." (Demonstrate
   careful lowering of chair, but stand by in case
   someone lets it down too fast.) "You are pulling
   down on the rope to lift the chair up. Does the pulley make the lifting job easier?"

Two proud boys, a rope, and a pulley lift and hold a
school chair at ceiling level.

Pass the cord over the pulley
  wheel; hang pulley on screw hook.

Group Discussion: (Try to find pictures of the following things to show.) "Pulleys help people load big ships, trains, and barns. Pulleys are used on steam shovels, on sailboats, on flagpoles, on scaffoldings, in draw-drapery rods, and inside the casings of windows. A pulley arrangement moves people up in large buildings on elevators and escalators. Single wheels help people and machines pull *down* to *lift up* or *pull across*. David Burnie's *Machines and How They Work* is an excellent source for pictures.

PRIMARY INTEGRATING ACTIVITIES

Math Experiences

Forming Sets. Use plastic cars (sold in bags at variety stores) and macaroni wheels as members of sets to link simple machine study with math activities.

Car Classification Game. Let children park tiny plastic cars in the correct parking area according to color. Use red, yellow, and blue construction paper as the parking garages where the cars of each corresponding color are parked. Add a second dimension to the classification so that red cars with tops are parked on one side of the red garage, red cars without tops are parked on the other side of the red garage, and so on.

Wheel Quantity Classification Game. Cut out and mount catalog pictures of wheeled objects. Ask children to sort them into groups according to the number of wheels each object has. (A record player turntable could be used as an example of a one-wheeled object.)

Music (Resources in Appendix A)

1. Sing this song about friction to the tune of "Here We Go Round the Mulberry Bush" while rubbing hands together.

Friction Song

This is the way we warm our hands, warm our hands, warm our hands.
This is the way we warm our hands, out in chilly weather.
Friction is what warms our hands, warms our hands, warms our hands.
Friction is what warms our hands, rubbing them together.

—J. H.

2. Beatrice Landeck's song "Little Red Wagon" is a lively reminder of the usefulness of wheels and axles. One child created these three verses for his favorite song:

> *"Both wheels are off and back end's dragging . . ."*
> *"Four wheels are off and we're not even moving . . ."*
> *"Now we're going to the repair shop . . ."*

3. Listen to Tom Paxton sing "Ride My Bike" on his cassette, *Suzy Is a Rocker*.

Stories and Resources for Children

ALEXANDER, ALLISON, & BOWER, SUSIE. *Power Magic.* New York: Simon & Schuster, 1991. Includes directions for making and using a pulley to hoist stuff upstairs where it probably belongs.

ARDLEY, NEIL. *Science Book of Machines.* San Diego: Harcourt Brace Jovanovich, 1992. The simple machine projects in this book are too time-consuming for the classroom, but the photographs clarify concepts well, and may inspire older children to experiment at home.

BURNIE, DAVID. *Machines and How They Work.* New York: Dorling Kindersley, 1991. Excellent color photographs and intricately detailed drawings allow children a clear view of wheels and gears in use in a clock, windmill, and 10-speed bike.

DOUGLASS, BARBARA. *The Great Town and Country Bicycle Balloon Chase.* New York: Lothrop, Lee & Shepard, 1984. Bicycles of all descriptions are seen in the great illustrations of this hot air balloon story.

GIBBONS, GAIL. *The Pottery Place.* San Diego: Harcourt Brace, 1987. The author contributes excellent illustrations of a foot-powered potter's wheel in action in this book.

HORVATIC, ANNE. *Simple Machines.* New York: Dutton, 1989. Fine photographs illustrate familiar objects that use simple machine principles. The clear text is suitable for early readers and pre-reading listeners.*

KONIGSBURG, E. L. *Samuel Todd's Book of Great Inventions.* New York: Atheneum, 1991. Delightfully illustrated by the author, this book narrates a day of noticing, appreciating, and wondering about the commonplace inventions that make a boy's life easier. It stimulates creative thinking and heightens interest in learning more about how the world works.

LAFFERTY, PETER, & JEFFERIES, D. *Pedal Power.* New York: Franklin Watts, 1990. Written for advanced readers, this history of the bicycle provides fascinating illustrations of gears and wheels for all ages.

MURPHY, BRYAN. *Experiment with Movement.* Minneapolis: Lerner Publishing, 1991. Brief descriptive sections on gears, levers, and friction provide good close-up illustrations.

OLLERENSHAW, CHRIS, & TRIGGS, PAT. *Gears.* Milwaukee: Gareth Stevens, 1994. The photographs of gears in use will increase awareness of their importance. Most of the activities seem too labor-intensive for classroom use, but may work as independent projects.

REY, H. A. *Curious George Rides a Bike.* New York: Scholastic Services, 1973. A hammer is used as a lever to open a crate. Gears, chain, and pedals on George's bike, and pulleys on a circus wagon cage are shown in this adventure.

SULLIVAN, GEORGE. *In-Line Skating.* New York: Dutton, 1993. Written for advanced readers, it illustrates the maneuverability of single wheels as the basis for this popular sport.

WILLIAMS, BRYAN. *On the Move.* New York: Random House, 1992. This small, clearly illustrated book on transportation includes drawings of bicycle gears, rollers moving airport baggage, rollers on paving equipment, rollers in the car wash, and wheels from unicycles to double-wide race car wheels.

WYLER, ROSE. *Science Fun with Toy Cars and Trucks.* New York: Simon & Schuster, 1988. Simple text for young readers. Toys are the basic equipment for experiments with ramps, friction, wheels, and the laws of motion. An important experiment demonstrates the purpose of seat belt use. Paperback.

ZUBROWSKI, BERNIE. *Wheels at Work.* New York: Morrow, 1986. This Boston Children's Museum activity book guides the building of simple wheel devices with inexpensive materials.

---

*Starred references, written at the young child's level of understanding, can help teachers who have minimal backgrounds in science to expand their knowledge base.

Poems (Resources in Appendix 1)

To stimulate children to invent machines they may wish existed, read "Homework Machine" from *A Light in the Attic* by Shel Silverstein.

Fingerplays

The familiar fingerplay, "The Wheels of the Bus," calls for arm rolling and active bouncing.

Art Activities

Collage. These collage materials are strongly suggestive of simple machines: macaroni wheels; pieces of thick string; bottle caps; Popsicle sticks; and round, triangular, and rectangular construction paper shapes. (The macaroni wheels may become essential parts of pulleys and gears, or they may be quietly eaten by a child whose inspiration is not as strong as his curiosity about new tastes.)

Cutting and Pasting. Young children enjoy making their own books or contributing pages to a group book for classroom use. For individual books, staple a few sheets of folded paper together. For group books, use cardboard punched with holes at the top. Fasten with leather thongs or notebook rings.

Provide scissors, paste, and pages cut from catalogs illustrating things with simple machine parts: clocks, eggbeaters, wheelbarrows, wheeled toys, mechanical toys, tools, carts, pepper mills, and so forth. If the supply of pictures is sufficient, they could be part of a math experience, making separate pages to show one-wheeled objects, two-wheeled objects, and so on.

Friction as an Art Medium. Sometimes when children are drawing with pencils and crayons, or making chalk rubbings, remind them that friction is involved in making colors stay on the paper.

A Friction Project. This project requires at least two days. Let the children sand 3" × 5" (8 cm × 12 cm) pieces of scrap lumber to satin smoothness. Next , use markers to draw a design on paper to cut out and glue to the sanded wood. The teacher can coat the plaque and picture with clear varnish and allow at least a day of drying time. Insert a small screw eye in the top edge of the plaque for hanging. This makes an appreciated gift for parents.

Dramatic Play

Packing Carton Vehicles. Children usually need little more than grocery cartons, paper plate wheels, and paper fasteners to create play cars and trains. A real steering wheel from an auto salvage yard, an inner tube and tire pump, and some sets of discarded keys add fun to the play.

Elevator. A tall refrigerator packing carton can become an elevator with a door cut in one side and numerals marked above it to indicate floors. It can be part of block or housekeeping play as desired.

Pulley Fun. Make an elevator for a skyscraper. Ask children to build a three-sided block tower directly below a pulley. Fasten a milk carton elevator to one end of the pulley cord. Cut a door in one side of the carton and fill it with toy passengers.

Attach two small pulleys to opposite sides of the block play area, a few feet above the floor. Tie the handle of a small basket to the pulley cord and knot the cord into a continuous, taut loop running between the two pulleys. It can be an aerial tramway for toy passengers or a conveyor belt for block construction workers.

Other Indoor Fun. A toy conveyor belt ramp is sold commercially. It can be an airplane baggage loader or a piece of play farm equipment. The two rollers that move the belt are operated with a small crank.

Add a toy sand or water wheel to sandbox or water play arrangements. They are sold seasonally as beach toys.

Provide pieces of perforated hardboard and hardware odds and ends to attach to it: cupboard doorknobs and hinges, nuts, bolts, and washers. Short screwdrivers are easiest for children to control.

Many commercial toys with visible working parts extend the simple machine concepts in play. Examples include construction sets, gear sets, clocks, music boxes, and locks with visible working parts. A pulley-operated cable car kit is available by mail from Hearth Song (see Appendix A).

A broken hand-wound alarm clock is a fine source of firsthand information about gears and springs. Provide small screwdrivers such as those sold as sewing machine accessories, and let the children take apart the clock. **Safety Precaution:** Steel springs can be sharp, so adult supervision is required for the dismantling.

Supply the workbench with bottle caps, metal ends from food container tubes, and discarded spools to inspire the construction of wheel-, gear-, and dial-encrusted inventions.

Creative Movement

Mechanical Toys. Children enjoy acting out the stiff, jerking movements of windup toys. Offer to wind up one child toy at a time so that other children can guess what kind of toy they are watching, or wind up all the toys at once so they can all move to music. In addition, children can try to move like the Tin Woodsman in the story "The Wizard of Oz," both when he was too rusty to move and after friction was cut down by oil so that he could move.

Creative Thinking

Read Shel Silverstein's poem "Homework Machine." Then ask children to imagine a new machine that would help them in some way. "What would it do? What would it look like?" Read the E. L. Konigsburg book, *Samuel Todd's Book of Great Inventions.* Encourage children to draw a picture of their inventions. Collect them into a class book of great inventions. Or let children write about a simple machine they like, or have used, and make a "What We Have Found Out About Simple Machines" class report.

What if? Collect pictures to illustrate a family picnic outing: getting food ready, climbing into the car, and gathering fishpoles and playthings. Ask, "What if there weren't any friction anywhere one day? What would be different for this family on a picnic? Yes, the mother would have trouble opening the peanut butter jar to make the sandwiches, the children would skid and fall down trying to walk to the car, the baseball bat would slip out of the boy's hand when he hit the ball, and the fishpoles might slide out of their hands."

Food Experiences

Children feel very important when they turn a crank that contributes to good eating. Several simple machines can be used to prepare food in the classroom. Make molded cookies with a cookie press. Use a hand-cranked grinder to make graham cracker crumbs for unbaked cookies or to make toast crumbs for the birds. Use a rolling pin roller to make cut-out cookies and a gear-driven eggbeater to mix instant pudding. Grind wheatberries in a hand-cranked coffee mill.

Field Trips

1. Field trips to see simple machines in action can be as simple as a walk to the school kitchen. A well-timed visit might allow the children to see a food order being unloaded on a hand truck from a delivery truck. Hand trucks have one pair of wheels on an axle and use the lever principle to lift loads. The kitchen may have a swivel-wheeled utility cart and a large can opener that has a sharp wheel to cut metal, gears to turn the can, a crank that turns an axle, and a screw to clamp the opener to a table.

2. The school office may have an electronic typewriter that has wheels, gears, axles, and a roller inside. Ask permission to lift the top for children's visual inspection (no touching!). Examine the gears of the plastic correction tape reels and the ribbon cartridge. Find the corresponding gears inside the typewriter that mesh with and turn them.

3. A repairman working in the building could show children an array of tools and equipment that apply simple machine principles.

4. The nearest driveway could be the scene of an impressive sight if the teacher can demonstrate the screw or lever principle by jacking up her car. (Be sure to observe the safety precaution of blocking the wheels on the ground with a brick.)

   Safety Precaution: Be sure to check for safety hazards when planning the following field trips. The locations are not intended for general public use. Children should be well-supervised.

5. A visit to an auto repair shop can verify the usefulness of simple machine principles. Mechanics in small garages may use scooting platforms on caster wheels to get under cars. A mechanic might demonstrate the tire changer, using a crowbar as a lever to pry off the tire.

6. Many wheels, rollers, conveyor belts, and pulleys are used in the work areas of the post office. Each one also has a flag pole with a pulley to raise the flag.
7. Backstage supermarket operations include hanging meat from overhead rolling tracks, using sets of rollers enclosed in metal frames to slide cases of food from trucks to storerooms, and moving groceries on checkout counter conveyor belts.
8. An older retail store in your community might still use a hand-operated dumbwaiter to bring stock upstairs from a basement storage room. In some types of dumbwaiters, the pulleys and ropes are visible. See if there is an enclosed freight elevator near your school where the whole class could take a ride together. It's fun to do.

There are so many opportunities to see simple machine principles in action that it would be easy to overdo the effort. Don't risk wearing down children's interest by taking too many field trips.

SECONDARY INTEGRATION

Maintaining Concepts

There are countless opportunities to bring simple machine principles to light during everyday school activities. When we want things to move or spin or slide easily, we can mention the need to cut down friction with a few drops of oil. There are times when too little friction poses safety hazards for young children: shoeless children moving fast on slippery floors or unwary children crossing streets on rainy or icy days.

For a vivid application of the lever and a pair of wheels, let children use a hand truck to help trundle heavy sandbags to the sandbox. Encourage them to try moving the sandbags without the hand truck so that they will realize how much the hand truck helps them.

Simple machine concepts are part of workbench use. Examples include the friction of sanding and sawing, the lever action of pulling out nails with the hammer, and the interlocking screw threads that hold wood tightly in a vice or clamp.

Whenever possible, let children watch the repair and maintenance of classroom and playground equipment. Better still, enlist the help of a shy child or one who needs a chance to get positive attention and approval. One child who broke toys and mistreated other children to attract attention was able to give up these undesirable behaviors when he became our chief mechanic. He took such pride in manning the oilcan and tightening tricycle bolts that he became a good steward of school property and a friend to other children.

Connecting Concepts

Simple machines help overcome the effect of gravity's pull. This can be brought out by speaking about the weight of the objects being lifted or moved in terms of gravity's pull. The effect of magnetism could be used to discover what a strong axle or wheel frame is composed of or to learn what kind of hinge rusts and needs oiling when it works hard and squeaks from too much friction.

Simple machine principles are used to move boats on water. Easy examples are the levers called oars and the paddle wheels that power the swanboats seen in Robert McCloskey's book, *Make Way for Ducklings.*

Family Involvement

Ask families if they would be willing to lend equipment or demonstrate skills that involve simple machines. Todd's mother loaned us a butter churn that her grandmother had used. Jenny's mother made bread with us, using her handcranked kneading bucket. Both children gained new status as a result of their parents' contributions.

RESOURCES

MACAULAY, DAVID. *The Way Things Work: A Visual Guide to the World of Machines.* Boston: Houghton Mifflin, 1988. This book gives the teacher the same revelations of "so that's how it works" that children gain from their study of simple machines.

VAN CLEAVES, JANICE. *Machines.* New York: Wiley, 1993. The projects and explanations provide additional experiences with pulleys, on inclined plane, the wheel and axle of a pencil sharpener, and more.*

WIESE, JIM. *Rollercoaster Science.* New York: Wiley, 1994. This book for advanced level students can be used selectively as a fascinating illustrated reference for simple machines, light, and gravity principles at work in the amusement park.

Teaching Resource

Moving Machines. Videotape. Bo Peep Productions, P.O. Box 982, Eureka, MT 59917. Parallels are shown between heavy construction machines at work and young children playing with toys like them.

14

Sound

The discovery that sound occurs when something vibrates can help children overcome fears about scary noises. Vibrations cause sound whether they originate in the distant clouds or in our own throats. The following concepts are explored in this chapter:

- Sounds are made when something vibrates.
- Sound travels through many things.
- Different sizes of vibrating objects make different sounds.

The experiences begin with a group activity to clarify the term *vibration* and to establish the idea that vibrations cause sounds. This is followed by experiencing vibrations, experimenting with mediums through which sound travels, and relating the pitch of sound to the size of vibrating objects.

Introduction. Help children understand what a vibration is by producing one. Say something like this: "Can you lift your arm and let your hand dangle from your wrist? Now, shake your hand as fast as you can. What do you call what is happening to your hand? Shaking? Wiggling? Jiggling? Wobbling? Those are good words. *Vibration* is another word that means moving back and forth very fast. Does your hand look different when it is vibrating? Some vibrating things move back and forth so fast that they look blurry. Another thing happens when something vibrates. You can find out if you listen very quietly. Hold your hand near your ear, then vibrate your fingers and hand very fast. What happens? Does someone hear a soft, whirring sound?

"Now put your fingers very lightly on the front of your throat, near the bottom. Very softly sing a sound like *eeeeee*. Do you feel something with your fingertips? Try it again. Did something inside your throat vibrate? Yes, you made a sound in there. Whenever we hear a sound, it is being made by something vibrating. Try touching your lips while you say *hum-mmmmm*. What's happening?"

CONCEPT: Sounds are made when something vibrates.

1. *Can we see things vibrate?*

LEARNING OBJECTIVE: To learn the meaning of the term *vibration* by creating and watching vibration.

MATERIALS:

Binding strip from vinyl
   folder cover, or flexible ruler

Coffee can with plastic lid, or
   sturdy drum

Sand, rice, or foam packing
   chips

Quart milk carton

Rubber bands

GETTING READY:

Cut window opening in one
   side of milk carton.

Stretch rubber band around
   length of milk carton.

SMALL-GROUP ACTIVITY:

1. "Watch and listen to this." Extend folder strip
   from edge of table or chair, like a diving board.
   Bend free end down, then release. "What did you
   see and hear?"
2. Show the milk carton. "What will happen if you
   pluck the rubber band? Watch and listen. You try
   it."
3. "It was easy to see the plastic strip and the rubber
   band vibrate. Some vibrations are hard to see.
   Let's put light things on this can lid to find out if
   the lid vibrates." Put sand, rice, or chips on lid.
   Tap the lid. "What happens? Now you try it."

2. Can we feel things vibrate?

LEARNING OBJECTIVE: To learn the meaning of the term *vibration* by creating and feeling vibration.

MATERIALS:

One or more of these: windup
   alarm clock, kitchen timer,
   toy music box

Combs

Waxed paper (tissue paper
   disintegrates when damp)

GETTING READY:

Cut pieces of waxed paper to
   fold over teeth of combs.

SMALL-GROUP ACTIVITY:

1. Pass unwound clock, timer, and music box among
   the children. "Do you feel these things vibrating
   now? Are they making sounds?"
2. Wind the equipment and pass again. "Now what
   do you feel and hear?"
3. "Try making some music with vibrating air and
   paper. Hold the paper at the bottom of the comb
   lightly between your lips. Now try to hum and
   blow a tune at the same time. The vibrations feel
   funny, but the sound is nice."
4. Tour the room and halls to feel vibrating electric
   motors (aquarium water pump, sound system
   speaker, water cooler, etc.).

Read *Whistle for Willie* by Ezra Jack Keats, a story about how to pucker the lips tightly
to blow out another kind of vibration.

CONCEPT: Sound travels through many things.

1. Does air carry sound?

LEARNING OBJECTIVE: To become aware that sound vibrations can make air vibrate to carry sound.

MATERIALS:

12" (30 cm) pieces of plastic garden hose

Plastic golf club cover tubes (very inexpensive at sports stores)

SMALL-GROUP ACTIVITY:

1. "What could be inside the pieces of hose? Put one end next to your mouth; put the other next to your ear. Whisper your name into the hose. Could you hear yourself? What could be vibrating inside the tube?"
2. "You can *feel* what is vibrating if you put your hand at the end of the hose. Try saying words like *toot* and *tut*. Can you feel something pushing on your hand each time you say a word?" (Air carried vibrations from their throats through the tubes.)
3. Let seated children enjoy speaking to each other through the long golf club tubes.

Read *Goggles* by Ezra Jack Keats.

Group Discussion: "Peek into the ear of the child sitting next to you. Do you see anything inside the ear that is vibrating? Could something invisible be in there?"

Recall how the sand or foam chips bounced around when someone tapped the coffee can lid. One vibrating thing made the things next to it vibrate. "Air vibrates when something next to it vibrates. That is the way sound is carried by air. It is the way sounds usually come to our ears. The air inside our ears also vibrates so we hear the sounds." Let children experiment with this idea by covering and uncovering their ears with their hands as they listen to some music. Could the air inside their ears vibrate very well when their hands were covering them?

Safety Note: "Our ears have an outside part that we can see and an inside part that we can't see. The inside part is so delicate that vibrating air can vibrate it. That is how we hear. We must be very careful with our ears to keep that delicate part working well. Can you think of some good safety rules for our ears?"

1. Does water carry sound?

LEARNING OBJECTIVE: To become aware that sound vibrations can make water vibrate to carry sound.

MATERIALS:

Two similar containers, such as plastic buckets

Pair of blunt scissors

Sponge, plastic sheet, or newspaper for spills

Golf club tube cut into 12" (30 cm) lengths

GETTING READY:

Half fill one container with water

SMALL-GROUP ACTIVITY:

1. "Take turns pressing one ear to the side of the bucket to listen to the sound I'll make." Open and shut scissors to make a steady clicking noise while holding them inside the empty bucket.
2. "I'll make the same sound in the water in this bucket. Listen the same way here. Listen for clicks. Are they louder or softer in the water? Put the golf tube in the water, then put your ear at the other end to listen. Is the sound louder coming through the air or through water?" Let the children make the scissor clicks for each other.

Note: If children in your class have gone fishing, they will understand now why people try to be very quiet when they are fishing.

3. Do solid things carry sound?

LEARNING OBJECTIVE: To become aware that sound vibrations can make solid objects vibrate to carry sound.

MATERIALS:

Table

Bottle caps, sticks, small rock, or anything that will make a noise on the table

Desirable: clock, aquarium pump motor, musical toy works

SMALL-GROUP ACTIVITY:

1. "We're going to make sounds for each other. Some may be very soft, so we'll have to listen closely." Tap one fingernail on the tabletop. "Does this sound seem loud? Now cover one ear with your hand and press the other ear on the table top while I do the same thing again."
2. "Which way was the sound louder, through the air or through the wooden table? Now put your ears to the table and close your eyes. Each one of you can take a turn tapping on the table with these things. We'll try to guess what made the sound."
3. Later compare sounds made by the clock, motor, or music works when held in the hand, placed on the table while heads are up, and on the table while ears are pressed to the table. Which sounds loudest? Finding out takes time. Keep the learning pace relaxed. Listen to the children's ideas.

4. Will string carry sound?

LEARNING OBJECTIVE: To become aware that sound vibrations can make taut string vibrate to carry sound.

MATERIALS:

Light string

Metal spoons

Bar of soap

Empty frozen juice cans, 2 for each telephone

1 1/2" (4 cm) nails

Hammer

GETTING READY:

Cut string into 2 1/2' (75 cm) lengths for spoon chimes.

Make a hole in center of each can bottom.

Cut string into 5' (1.5 m) lengths for metal can telephones. (String can be longer for use at home. Short lengths are best when many children will be using phones in the same room!)

SMALL-GROUP ACTIVITY:

1. Dangle short string over a finger. "Do you think loose string will vibrate to make a sound if you pluck it? What if it is held tightly by both hands? Try it both ways with your string." (Loose string vibrates too slowly for audible sound.)
2. "Let's add heavy vibrating metal to the string ends and listen to find out if tight string will carry sound." Tie a spoon to each end of string. Fold string in half; press folded end to ear. Lean over so that spoons dangle freely. Swing string to make spoons strike each other. Listen to the sounds traveling up the string.
3. Make metal can telephones along with children. To stiffen string ends, pull them across soap bar. Put a string through the holes in two cans from the outside. Pull string into the can and tie around a nail. Wedge the nail inside the can. "Keep the string straight and tight. You talk into one can and your friend listens through the other can."

Vibration turns ordinary string and ordinary spoon into sound.

Group Discussion: Give each child a rubber band to pluck when it is limp and when it is stretched. What did the spoon chime and metal-can telephone makers find out about how string vibrates to carry sound? How many things vibrated to carry the sound of their voices? (Air in cans, the cans, and the string.)

CONCEPT: Different sizes of vibrating objects make different sounds.

1. Do different vibrating string lengths sound alike?

LEARNING OBJECTIVE: To hear pitch differences when different string lengths vibrate.

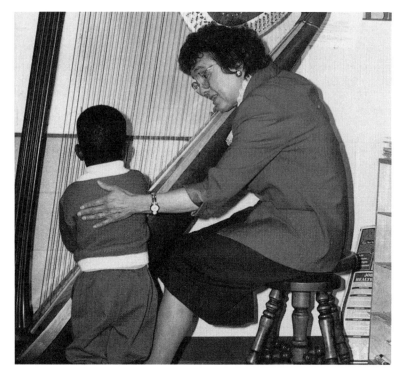

Vibrating strings, long and short, captivate Steven.

MATERIALS:

Three sizes of plastic boxes, such as 3", 4", 5" (8 cm, 10 cm, 13 cm)

Three sizes of small rubber bands

Empty covered boxes like a 2-pound (908 g) cheese box or small shoe boxes

Rubber bands large enough to stretch around the length of covered boxes

Small blocks

Autoharp or guitar (desirable)

GETTING READY:

Stretch small rubber band over each opened plastic box, the smallest band on smallest box, etc. (The effect won't be the same if same

SMALL-GROUP ACTIVITY:

1. Pluck rubber bands on the plastic boxes. "Do these all sound alike? Which one sounds highest? Lowest? Try it."
2. "Now try this." Stretch long rubber bands around lengths of covered boxes. Slip block under band near left end of the box. Pluck the length of rubber band to the right of the block. Listen. Continue plucking and slide the block slowly toward the other end of the box. "Is pitch changing? Is it higher or lower? Does the whole rubber band length vibrate, or just the side being plucked? Is the part that vibrates getting shorter and higher?" Let children experiment, sliding the block and plucking. "Can you play the scale this way? Play a tune?"

size rubber band is used on
all three boxes. Try to find
out why.)

Group Discussion: The autoharp very clearly shows the relationship between string length
and pitch. (A real harp and an upright piano with the front panel removed, or a grand piano with the top raised, will show the same thing.) If a guitar is used for this purpose, be
sure to make clear which part of the string is vibrating, so that the relationship between vibrating string length and pitch can be understood.

2. Do different vibrating air column lengths sound alike?

LEARNING OBJECTIVE: To hear pitch differences when different air column lengths vibrate.

MATERIALS:

Carton of 8 empty 16 oz. (480
ml) bottles

Pitcher of water

Funnel

Masking tape

Sponge

Bicycle tire pump or bellows
step pump

GETTING READY:

Experiment at home with
levels of water needed to
produce tones of the octave.
Internal shapes of bottles
vary, but these water levels
may be about right:
1. empty

2. 1 3/4" (4.5 cm)

3. 2 1/2" (6.5 cm)

4. 4 " (10 cm)

5. 4 1/2" (11.5 cm)

6. 5 1/4" (13 cm)

7. 5 1/2" (14 cm)

8. 6" (15 cm)

Place band of tape around
each bottle, with upper
edge of tape at measure-
ment level.

Let children pour water to
marked levels of bottles, if
possible.

SMALL-GROUP ACTIVITY:

1. "What is inside these bottles? What will happen if
   the air inside the bottles vibrates? When air moves
   across the top of the bottle, the air inside vibrates.
   Listen." Have a child use the pump while you direct a stream of air from the hose *straight* across the
   bottle top.
2. "We put some water in these bottles. Which bottle
   has space for more air? Let's see what happens
   when we make different amounts of air space vibrate."
3. Arrange bottles in order, 1 through 8. While a child
   pumps air, direct air from the hose across tops of
   bottles to play the scale. Try to play "Yankee
   Doodle." Let children play your bottle organ. (Put
   bottles in two cartons to prevent breaking, or keep
   bottles on the floor.)

It takes cooperative effort to make water organ music.

Note: Listen to a beautiful recording of Romanian pan flute music. Zamfir plays this ancient pan-pipes instrument. The recording cover illustrates the graduated-size pipes he blows across.

PRIMARY INTEGRATING ACTIVITIES

Math Experiences

Ordering by Pitch and Size. The mathematical relationship between the size and the pitch of vibrating objects is fun to explore. Sets of bells and sets of detached metal chimes, available commercially, can be placed on a molded framework to form a xylophone. The bells and chimes can be arranged by children in one-step tone order visually according to size, largest to smallest; or according to tone, lowest to highest.

Older children can have fun converting the bottle organ (p. 247) into bottle chimes when they tap the bottles lightly with a pencil. They discover that the tone order is reversed. The bottle with little air space and a large amount of water produces a high tone when the air vibrates, and a low tone when the water vibrates.

Because young children may be confused by this reversal, it would be wiser not to present this activity to them in two different ways.

Music

Many familiar children's songs are easy extensions of the learning about vibrations and pitch as they describe bell sounds and animal sounds. The concepts can be applied when rhythm instruments are used. "Charles, try holding your triangle by the string loop instead of with your hand. See if it vibrates more now." Prolong the life of school drums by suggesting that they will vibrate better when lightly tapped than when thumped very hard.

Children can make simple instruments to play. Directions are offered in *Making Music: Shake, Rattle and Roll with Instruments You Make Yourself,* by M. Jackson.

The following song is about the cause of sounds:

Sounds

—J.H.

Stories and Resources for Children

ARDLEY, NEIL. *The Science Book of Sound.* San Diego: Harcourt Brace Jovanovich, 1991. Lively color photographs illustrate directions for a variety of sound activities. A paper noisemaker toy suggests how a thunderclap is heard. A Reading Rainbow selection.

BERGER, MELVIN. *All About Sound.* New York: Scholastic, 1994. Simple activities are offered, together with explanations of vocal cord functioning, and how a stethoscope works. Paperback.*

BOOTH, BARBARA. *Mandy.* New York: Lothrop, Lee & Shepard, 1991. Noticing that things around her are vibrating lets a profoundly deaf child infer that sounds are being made.

BRETT, JAN. *Berlioz the Bear.* New York: G. P. Putnam's Sons, 1991. A buzzing bee causes one problem for the village band and solves another. Rich detail and whimsy of the author's illustrations add visual delight to this Reading Rainbow selection.

CUMBAA, KAREN, & S. *Bones and Skeleton Game Book.* New York: Workman, 1993. Includes an illustrated description of how sounds reach hearing centers in the brain, and offers an experiment to sense sound vibrations going through the bones in the head.

*Starred references, written at the young child's level of understanding, can help teachers who have minimal backgrounds in science to expand their knowledge base.

EVANS, DAVID, & WILLIAMS, C. *Sound and Music.* New York: Dorling Kindersley, 1993. Bright color photographs of children engaged in simple explorations of sound making will lead young story listeners into experimenting.

JACKSON, MICHAEL. *Making Music: Shake, Rattle and Roll with Instruments You Make Yourself.* New York: Harper Collins, 1993. Offers directions for making and playing no-cost instruments, as well as instructions for playing the harmonica. Paperback.

JENNINGS, TERRY. *Sound and Light.* New York: Smithmark, 1992. Interesting, simple sound experiences and sound facts make up this book, including sketches of the human voice mechanisms and the inner ear.

JEUNESSE, GALLIMARD, & DELAFUSSE, C. *Musical Instruments.* New York: Scholastic, 1992. Color illustrations and simple information about the instruments of the orchestra clarify the relationship between the length of vibrating strings or air columns and the pitch they produce.

KANER, ETTA. *Sound Science.* Reading, MA: Addison-Wesley, 1991. This light-hearted compendium of fascinating sound facts, riddles, and activities for older children can provide fresh ideas for extending sound studies for young explorers.

KEATS, EZRA JACK. *Goggles.* New York: Collier-Macmillan, 1971. Peter sends his voice through an empty drainpipe to confuse the big boys who are trying to catch him.

KEATS, EZRA JACK. *Apt. 3.* Englewood Cliffs, NJ: Prentice Hall, 1986. Varied sounds guide Sam to a sightless new friend who knows his neighbors by the sounds he associates with them. Paperback.

LILLEGARD, DEE. *Strings: An Introduction to Musical Instruments.* Chicago: Childrens Press, 1988. Photographs and illustrations of stringed instruments, with a brief explanation of just what vibrates when the instruments are played. Also in this series: *Brass* and *Woodwinds.*

PINKNEY, BRIAN. *Max Found Two Sticks.* New York: Simon & Schuster, 1994. Max captures the joy and rhythm of the sounds of life around him with his two special sticks.

POLLACO, PATRICIA. *Thunder Cake.* New York: Philomel, 1990. A young girl is distracted from her fear of thunder in this appealing story.

TAYLOR, BARBARA. *Hear, Hear: The Science of Sound.* New York: Random House, 1991. Sound experiments and activities are offered with brief explanations of underlying principles.

WHITE, LAURENCE. *Science Toys and Tricks.* New York: Harper & Row, 1989. Some simple sound experiments are presented in this collection of activities. One lets you listen to a fly walk (but first you have to catch a fly!).

Poems (Resources in Appendix A)

Many poems for children recreate sounds and extend the learning activities about sound.

> From *Poems To Grow On* by Jean McKee Thompson, read "Kitchen Tunes" by Ida Pardue, and "The Storm" by Dorothy Aldis.
> From *Poems Children Will Sit Still For*, read "Wind Song" by Lillian Moore.
> Paul Fleischman's *Joyful Noise* captures the sounds of familiar insects.

Fingerplays

This fingerplay relates size to pitch and volume:

The Clocks

(slowly, loudly)
When I'm a big tall clock,   (stand & stretch tall)
I go TICK TOCK, TICK TOCK, TICK.   (swing arms like a pendulum)

(faster, softer)
When I'm a middle-size clock
I go tick tock, tick tock, tick.   (stoop over)
                                                          (swing arms faster)

(even faster, softer)
When I'm a very small clock   (crouch low)
I go tick-tick-tick-tick-tick   (move hands fast)
So fast that I fall over!

                                                          —Author unknown

Recall the metal-can telephone fun with this fingerplay.

      Tin-Can Telephone

I called my friend on a tin-can
  phone.   (Make cylinders with hands)

Two tin cans were joined by a
  string.   (extend little fingers to touch each
                                                            other)

I put it to my mouth,   (put one cylinder up to mouth)
I put it to my ear,   (put other cylinder to ear)
I could hear my friend.

                                                          —J. H.

Art Activities

Tambourines. Let children make crayon designs on two paper plates. Punch holes at regular intervals around the edges of the pair of plates. Children can lace the plates together partway, fill the tambourine with bottle caps, then complete the lacing. Plastic margarine tubs can also be used, but they are difficult to individualize with the child's own designs.

Drums. Use salt or oatmeal boxes, or cans with air-tight plastic lids. Glue construction paper to the side of the drum, slipping rubber bands around the paper to hold it firmly while the glue dries. Children can decorate with crayons.

Horns. Let children wind and glue lengths of yarn around paper towel tubes. (The pressure children use when crayoning will squash most tubes.) An adult can punch an air-vent hole near one end. Help children fasten a single piece of waxed paper over one end of the tube with a tight rubber band. A reedy sound is made by tooting tunes into the open end of the tube.

Jingle Sticks. For each stick, use a hammer and thick nail to make a hole through the center of six bottle caps. Let children sand 8″ (20 cm) lengths of dowel, 1 1/4″ (4 cm) in diameter. Secure the dowel in a clamp or vise. Let children use roofing nails to loosely attach three pairs of bottle caps (open edges together) near one end of the dowel.

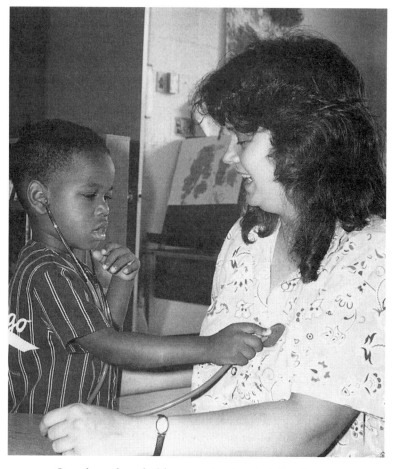

Joseph can hear hidden sounds when he's the doctor.

Dramatic Play

1. A pair of tin-can telephones can add interest to the block corner airplane, ship, or spaceship play. They can also be fun to use in a pretend office, home, or hospital.

2. Medical play can include the experience of actually hearing each other's heartbeats. Use a real or improvised stethoscope to "gather" the sound so it can be heard more clearly. Insert the tube end of a soft plastic funnel into a short length of narrow-gauge plastic tubing. Place the funnel on a child's chest. Hold the tubing end close to *(not in)* the ear to hear the beats.

Creative Movement

Children can show their perceptions of high or low sounds by moving to the tones of an instrument played by an adult. When the notes are low the children move in a low position close to the floor; when they are high the children move in a stretched-up position. They can also rise from a crouch to tiptoes in response to a slowly ascending scale, and return to the crouch as the descending scale is played.

Creative Thinking

Match My Sound. Gather pairs of matching items that will produce sounds, such as metal spring "crickets," two squares of sandpaper to rub together, small brass bells, jingle bells, two bottle caps to tap together, blunt scissors to click, small pieces of corrugated paper to scrape with a fingertip, pocket combs and waxed paper to blow on, two finger cymbals. Put one of each item in a teacher's bag. Distribute the matching soundmakers as evenly as possible into lunch-size paper bags. Fold down the tops and present a bag of secret sounds to each child at the table. While children keep their eyes tightly shut, the teacher reaches into her bag and makes a sound with one of the items. The children then search through their bags to see who can find things that produce a matching sound. Allow plenty of time for this game. The children may want to trade bags and begin again when you finish.

Sound Matching Tray. Use a divided cutlery tray and color film cans or small opaque plastic bottles to adapt a Montessori sensory training exercise. Loosely fill two matching containers with materials that make distinctive sounds when shaken, such as gravel, sand, pennies, rice, or puffed cereal. Mark matching pairs with matching numerals or letters on the container bottom as a self-correcting aid. Put one set of containers on one side of the tray; have children find the matching sound containers and place them next to their mates on the other side of the tray.

Thinking Together. Make lists of sounds that help keep us safe, sounds we enjoy hearing, and sounds that can be scary.

Field Trips

Look around your school area for a place that will produce an echo. An empty gymnasium or a high, windowless wall that borders an empty lot might be good places. Tell children that when they shout their names toward the walls, the vibrations touch the walls and bounce back to them. Try it.

As you walk together for any special purpose, suggest walking very quietly so that you can listen for sounds to identify. An old church or auditorium with visible organ pipes, within walking distance of the school, is a good destination for this study.

SECONDARY INTEGRATION

Maintaining Concepts

Occasionally two of the loudest and most frightening sounds occur while children are in school: thunder and sonic booms. When lightning moves through the air, that air suddenly gets warm and it swells so fast that it vibrates. Thunder is the way we hear the vibration from that fast-swelling air far up in the clouds.

The unexpectedness of sonic booms adds to the scariness of that particular sound. Adults can agree with frightened children that surprise noises startle us. That acknowledgment and the example of our calm response are helpful to children. You could explain that when a jet goes very, very fast, it sometimes bumps into the air ahead that is already vibrating with its own sound. Sound travels fast, but sometimes a jet can travel faster than sound. People on the ground below hear that bump like a big explosion of air.

Connecting Concepts

Some of the suggested experiences relate to the children's existing concepts about the presence of air in empty-looking places and moving air pushing things (see Chapter 7, What Is Air?). Friction is also involved in the production of some sounds, such as rubbing sandpaper together and unlubricated metal parts rubbing together to produce squeaks from the vibrations (see Chapter 13, Simple Machines).

Family Involvement

Any parent who is able to bring in and play a musical instrument can enrich the children's awareness of soundmaking. Families could be encouraged to lend things like a bell collection or wind chimes for the children to enjoy. Suggest events of soundmaking interest in the locality that parents might visit with their children, such as a bell-ringing choir, a pipe-organ recital, or concerts.

RESOURCES

HANN, JUDITH. *How Science Works.* Pleasantville, NY: Reader's Digest, 1991. Excellent background how and why information about sound, including the difference between sound and music.

SELLER, MICK. *Sound, Noise and Music.* New York: Glouster Press, 1992. Good background information about sound as a form of energy, how it travels and is heard, and how music is made.

Light

On the day of birth, infants are able to perceive light. Later, the playpen explorer may try to catch a sunbeam. The toddler may try to pounce on a shadow. Growing young children are fascinated by the sparkle of reflected light and the beauty of the spectrum. The experiences that capture some children's closest attention, however, are those that help to allay their worries about darkness—the absence of light. The following concepts are suggested for investigation in this chapter:

- Nothing can be seen without light.
- Light appears to travel in a straight line.
- Shadows are made when light beams are blocked.
- Night is Earth's shadow.
- Everything we see reflects some light.
- Light is a mixture of many colors.
- Bending light beams make things look different.

Children can experiment by examining a box full of darkness, using a flashlight beam to note the straight path of light, creating shadows, and looking through filters. Other experiences explain night and day, reflection, and refraction.

Introduction: To lead into these experiences with light, read one of these stories about the mastery of nighttime anxiety: *Bedtime for Frances* by Russell Hoban or *Switch on the Night* by Ray Bradbury.

CONCEPT: Nothing can be seen without light.

1. What can we see in a dark box?

LEARNING OBJECTIVE: To become aware that nothing can be seen without light.

MATERIALS:

Pen flashlight

Extra batteries

Shoe box with cover

Small picture

Old heavy blanket

Low table

GETTING READY:

Cut a dime-size peephole in
one end of the shoe box.

Cut a flap in top of box.

Paste picture to inside of box
at end opposite the peep-
hole.

Cover table with the blanket
so it drapes to floor on all
sides, making a dark place.

SMALL-GROUP ACTIVITY:

1. "Let's stretch out on the floor and put just our
heads into the dark place under the table."
2. "Here is a dark box. It has a hole in one end to look
into. What do you see in there?" Pass the box to
everyone.
3. "Now I am going to change something; then you
can look into the box again." Turn on the flashlight
and push it under the flap on the top of the box.
"Can you see something in the box now? It was
there before, but you couldn't see it. Let's try to
find out why."
4. Remove the flashlight; pass the box again. "The
picture is still in there. Can you see it? What happens when the light goes off? We always need light
to see anything. *Nothing can be seen without light.*"

Be prepared to repeat this experience. Children are impressed with, and reassured by, their
ability to make the dark go away and come back at will.

CONCEPT: Light appears to travel in a straight line.

1. Does a flashlight beam curve around things?

LEARNING OBJECTIVE: To notice that light appears to travel in straight lines.

MATERIALS:

Pen flashlight

Extra batteries

3 sheets of white paper

Blanket and low table

Blocks

Masking tape

GETTING READY:

To prepare 3 screens:

Stack the 2 blocks and tape the
flashlight to the top block. Turn
the flashlight on.

Fold paper as shown in Figure
15–1. Stand it in front of the
light.

SMALL-GROUP ACTIVITY:

1. "Let's find out how a beam of light travels. Lie
down along the sides of the table. Put your
heads under the blanket. I'll be at this end; no
one will be at the other end."
2. Turn on the flashlight. "Notice the light shining
on the blanket at the end of the table? It passes
through each hole in the screens."
3. "Liz, will you put your hand in front of one
hole, please? What happened? Did Liz's hand
block the light or did it curve around the paper
to shine through the next holes?"
4. "Take turns blocking the light beam as Liz did.
Does the light curve around things or does light
travel in a straight path?" Let children use the
flashlight to check and recheck the path of light.

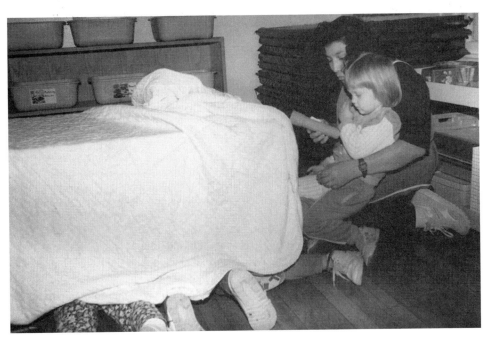

A blanket-covered table adds intrigue to light experiments.

Cut a small hole where the
 beam hits the center of the
 paper. Cut holes in the same
 location in the other screens.

Line up screens a few inches
 apart so that the light beam
 can pass through all the holes.
 Cover table with blanket.

CONCEPT: Shadows are made when light beams are blocked.

1. Does light shine through some things and not others?

LEARNING OBJECTIVE: To perceive light differently as it shines through different materials and to
                    become aware that shadows result from blocked light.

MATERIALS:

Small flashlight

Blanket and low table

Waxed paper

Cardboard

SMALL-GROUP ACTIVITY:

1. "Let's find out what light will shine through. You
   stretch out along that side of the table and I'll be
   across from you."
2. Put the box on its side with the opening facing the
   children. Shine the light through the opening.
   "Let's see if light shines through air."

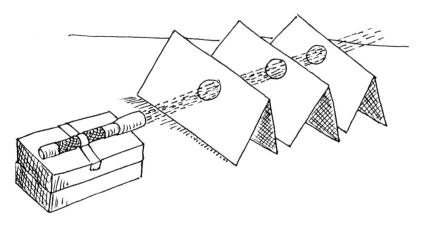

FIGURE 15-1

Large shoe box

Clear glass jar

Water

Colored tissue paper

Clear acetate folder

GETTING READY:

Cut a large opening in shoe
  box bottom.

Cut simple 3″ (7.5 cm) cardboard doll figure.

Fill jar with water.

Cover table with blanket.

3. Put the water jar in the opening. "Let's see if light
   shines through water."
4. "Let's see if light will shine through waxed paper
   in the same way." Cover the opening with waxed
   paper. "Does the light shine in one spot or spread
   out?" Try colored tissue paper and clear acetate at
   the opening.
5. "Let's see if light will shine through this cardboard doll." Remove box; place doll so that it
   makes a shadow on the blanket.
6. Tell children to notice that light shines on only
   one side of the doll. "There is a dark place where
   the doll blocked the light. It's a shadow."

Large-Group Activity: Move outdoors on a sunny day. Use chalk to trace the outline of children's shadows on a paved area. As you trace, ask other children, "Is sunlight shining right
through Nina, or does she block the light? I'm tracing the darkness that Nina's body makes
where the sunlight can't shine through her."

   Note: A child may notice that a transparent glass window sometimes *reflects* light like
a mirror. This occurs as a result of the way light strikes the surface of the glass. The surface
of a still pond or shallow swimming pool can also show a mirror-like reflection. This happens when sunlight reflects back *up* from the bottom of the pond or pool.

CONCEPT: **Night is Earth's shadow.**

1. Why do we have day and night?

LEARNING OBJECTIVE: To begin to understand night's darkness as the turning Earth's shadow.

Look, we're blocking the light!

Group Experience: (You will need a globe or ball, projector or flashlight, chalk, and a darkened room.) "We had fun blocking the sunlight to make shadows. Did you know that the whole Earth does the same thing? That is the way day and night happen. Let's pretend that this light is the shining sun. This globe is a model of our huge Earth. I'll put a chalk mark here on the globe to show where our part of the world is. Watch the mark as I turn the globe the way the Earth always slowly turns. Is the mark in the light now? Does the light curve around the globe to keep shining on our part of the world? No, the light shines in a straight line. Our part of the world is in a shadow now—the shadow of the other side of the Earth."

Continue to turn the globe, so that children can see the marked part of the globe alternately in the light and in the shadow. "What do we call the time when our part of the world is in the Earth's shadow? When it is in the sunlight?" Let the children turn the globe.

When your class encounters mention of the sun's *rising* or *setting* in stories, poems, or songs, point out that such was the old, imaginary way of thinking about night and day. *We* know that it only *looks* like the sun is traveling around the Earth. We have daylight when our part of the traveling Earth spins toward the sun. After a while, rely on the class to correct such misconceptions they may find.

CONCEPT: Everything we see reflects some light.

Introduction: Put a lighted purse-size flashlight into your cupped hands. Open your hands to let the children see where the light is coming from. "Is the flashlight making this light?" Hold a pocket mirror in your hands. "Is the mirror making a light?" Now shine the flashlight beam onto the mirror. Tilt the mirror so that the light is reflected on the walls or ceiling. "Which thing made the light that is bouncing up there, the flashlight or the mirror?"

Turn the flashlight off and on several times so children can verify that the spot of light is reflected by the mirror to the ceiling only when the flashlight makes the light.

1. Do some things reflect light better than others?

LEARNING OBJECTIVE: To become aware that various materials reflect different amounts of light.

MATERIALS:

Strong sunlight, or spotlight study lamp

Large sheet of stiff white paper or cardboard

Aluminum foil pans

Shiny baking sheets

Colored construction paper: yellow, red, black

Dark carpet sample, velveteen, or bath towel.

SMALL-GROUP ACTIVITY:

1. Prop white paper on a chair seat at a right angle to the light source.
2. "Can you stand facing the light, holding a pan so it reflects light to the sheet of paper on the chair? See the reflection glow? Try to reflect light on the paper with these other things. See which ones reflect light well and which ones do not." Start with things that reflect well. Children may be discouraged as they try things if they don't get results right away

Whole-Group Experience: To use aluminum foil conservatively, do this for all to observe: Reflect light onto a smooth, new sheet of aluminum foil. Then crumple the sheet and reflect light onto it, to compare the diffused reflection with the first reflection. (The cumpling created many small surfaces.) **Safety precaution:** Small pocket mirrors are fun to use as reflectors, but they may be hazardous. Tape edges of unframed mirrors if you decide to use them.

CONCEPT: **Light is a mixture of many colors.**

1. Can some things bend light to show its colors?

LEARNING OBJECTIVE: To observe the effect of light passing through a prism (or prism-like materials) and spreading into spectrum colors.

MATERIALS:

Magazine

Strong sunlight, or a projector

Small piece of thick plate glass or plastic, edges taped*

Prisms, or chandelier drops

Small, clear plastic boxes

Water

Soap bubble materials

Aquarium (desirable)

Diffraction grating cards, if available.

SMALL-GOURP ACTIVITY:

1. "There is a surprise in light beams. This magazine can help us find it." Hold the magazine; cut edges toward children, bound edge toward you (Figure 15–2). "When pages are pressed together, does the edge look like one thick line? I'll bend it back toward me. Notice the line spread out to show many edges of paper. Try it."
2. Hold plate glass in front of light source. "When light comes through this glass, it looks clear. Let's see how light looks when it is spread by glass that bends it."
3. Hold prism edge in front of light beam and rotate until a spectrum can be seen in the room. "Now

---

*Do *not* use a magnifying glass.

FIGURE 15–2

you see the secret! Light really has seven colors in it (spectrum), but we only see them when light shines through something clear that is bent or curved." Let the children try it.
4. If possible, place an aquarium near sunny windows. Fill with water. Help children stand where they can see the spectrum through the aquarium corners.
5. On a sunny day, let children experiment with water in small plastic boxes outdoors. Can they see the spectrum through the corners?
6. Try to make soap bubbles near a sunny window to see colors.
7. Let children examine diffraction grating cards to see the rainbow effects reflected by the patterns of lines.

CONCEPT: **Bending light beams make things look different.**

1. How do things look through a curved drop of water?

LEARNING OBJECTIVE: To observe that things change in appearance when the path of light bends as it travels through transparent, curved materials.

MATERIALS:

Small bottles

Water

SMALL-GROUP ACTIVITY:

1. "What happens when you put *just one* drop of water on your piece of waxed paper? Look at the

Medicine droppers

Waxed paper

Newspapers

Magnifying glasses or plastic "Card Lens" magnifiers

Sponges or paper towels

Small objects to examine

GETTING READY:

Fill bottles with water for each child.

Cut waxed paper and printed sections of newspaper into 3" (7.5 cm) squares.

Give each child a square of newspaper with waxed paper covering it.

newspaper through the drop of water. What do you see? Is the water drop curved or flat?"

2. "Now add more water to the drop. Is it still curved, or is it a flat spot of water? Do words beneath still look bigger? Why not? What has changed?"

3. "Absorb the water with a sponge and try again."

4. Show children the curved profile of a magnifying lens. "Hold up the curved water bottle. Look at your finger through the bottle and water. How does it look?" (The path of light bends when it goes through curved glass or a curved drop of water. Things seen through them look different.) "We looked through curves that made things look bigger." Look at the curved grooves in the plastic "Card Lens" magnifiers. Make the lenses and magnifying glasses freely available for children's own independent explorations from now on.

Note: Children are intrigued by the upside down reflections they see in the concave (inside) surface of a shiny spoon, or the odd magnification made by the convex (outer) surface of the spoon. Point out that the curved path of light changes reflections on these curved surfaces.

PRIMARY INTEGRATING ACTIVITIES

Math Experiences

Shadow Math. Early on a sunny morning move outdoors and have children find a place to stand along a line, with their backs to the sun. Let them trace one another's shadow outlines with chalk. Mark with the owner's initials. Measure the length of each shadow with a meter stick, and record the data indoors on a chart. At noon, return to the marked places and repeat the tracing and data collecting. While abstract explanations of the earth's spinning will not make much of an impression on most children, this experience with changing shadows will contribute to eventual understanding.

Return to the marked places at the end of the school day. Where are their shadows? Measure the length of the afternoon shadows. Add to the chart. (As our part of Earth spins away from the sun, shadows change in direction as the day goes on. They change in size as the angle of sunlight reaching the Earth changes.)

Make a Sundial. In late spring, older children will be intrigued by creating a primative sundial on the school grounds that can be closed off for this purpose. Good directions for this project can be found in Sandra Markle's activity book, *Exploring Summer.* (See Stories and Resources for Children.)

Mirror Math. Explore easy symmetry experiences with mirrors. Provide small, rectangular mirrors to children. Give them pictures cut from magazines to see if they can make the same pictures in the mirror. Provide objects like half circles to see if they can make whole circles in the mirror.

Further ideas for making these activities more challenging for older children are found in the Delta publication, *Mirror Explorations.* (See Appendix A.)

Energy Conservation. (Adding tens in a useful context.) Show children the wattage numbers on lightbulbs, explaining that the numbers mean different degrees of brightness. Using a small shadeless lamp, demonstrate the brightness of different wattages. Calculate with the children the total wattage used in the classroom, so they can understand the reason for turning off the lights when the classroom is empty.

Explore Angles. As a background experience for future geometry study, let children play with a flashlight and mirror to explore how the angle of light hitting the mirror is the same as the angle of light bouncing off the mirror. Pose problems for them, "Can you use the mirror to make the light shine on the door . . . on the ceiling . . . on the big table?"

Billy catches Rose's flashlight beam in his mirror. He bounces the reflection all around the room.

Music (Resources in Appendix A)

Mention that laser light beams are used instead of needles to play compact disc recordings. Sing this song after children have used prisms (to the tune of "Good Morning, Merry Sunshine"):

Prism Song

Good morning, merry sunshine,
Your light comes straight to me.
When glass or water bend your line,
Your colors I can see.

—J. H.

Stories and Resources for Children

ARDLEY, NEIL. *Great Science Experiments.* New York: Dorling Kindersley, 1993. This collection of science activities includes directions for making a periscope and a kaleidoscope.

ASIMOV, ISAAC. *Why Does the Moon Change Shape?* Milwaukee: Gareth Stevens, 1991. This focused explanation of the way we see the moon includes moonlight as reflected sunlight, and how the changing shapes of the moon are the changing shadows Earth casts on the moon.

BRADBURY, RAY. *Switch on the Night.* New York: Harper Collins, 1993. A child learns to control his fear of darkness. Read after exploring darkness as the absence of light. Paperback.

BRANLEY, FRANKLYN. *Eclipse: Darkness in Daytime.* New York: Harper & Row, 1988. Clear information for kindergarten/primary-grade level.

BRANLEY, FRANKLYN. *Light and Darkness.* New York: Thomas Y. Crowell, 1975. The difference between things that create light and those that reflect light is clearly shown. Reassuringly emphasizes the constancy of the sun's light.

BRANLEY, FRANKLYN. *What Makes Day and Night.* New York: Harper & Row, 1986. Earth's rotation is viewed from the vantage point of a spaceship, and a simple experiment at home personalizes what makes night and day. Paperback.

BROEKEL, RAY. *Experiments with Light.* Chicago: Childrens Press, 1986. Includes enrichment-level experiences with refraction and color, as well as basic information about light.[*]

BROEKEL, RAY. *Experiments with Water.* Chicago: Childrens Press, 1988. Two refraction experiments described in this book link water and light topics.

BULLA, CLYDE. *What Makes a Shadow.* New York: Harper Collins, 1993. A simple text informs preschoolers about shadows as blocked light paths, and night as Earth's shadow. Paperback.

CONAWAY, JUDITH. *More Science Secrets.* Mahwah, NJ: Troll, 1987. Directions for making a solar-powered pinwheel and "solar mittens" are included in this activities collection. Paperback.

CORBY, JANE (ED). *Shadowgraphs Anyone Can Make.* Philadelphia: Running Press, 1991. Instructions and movement suggestions for making 28 animal shadow shapes with two hands, ten fingers, and a light source.

---

[*]Starred references, written at the young child's level of understanding, can help teachers who have minimal backgrounds in science to expand their knowledge base.

DOROS, ARTHUR. *Me and My Shadow.* New York: Scholastic, 1990. Simple, clear explanations are presented for night as Earth's shadow, and the phases of the moon as its own shadow. Shadow tag and other shadow fun are included.

FLEISCHMAN, PAUL. *Shadowplay.* New York: Harper & Row, 1990. A morality tale within a story of children discovering how a shadow puppet play is created.

HOBAN, RUSSELL. *Bedtime for Frances.* New York: Harper & Row, 1960. Monstery shapes become familiar chairs and robes when father turns on the light in Frances' bedroom.

JENNINGS, TERRY. *Energy and Forces.* New York: Smithmark, 1992. A section on light energy from the sun offers directions for making a solar cooker.

JENNINGS, TERRY. *Sound and Light.* New York: Smithmark, 1992. The experiments for understanding and using light include making a very simple lens, and observing the effect of light on sow bugs and on plant growth.

LIONNI, LEO. *Little Blue and Little Yellow.* New York: Morrow, 1994. A color story that intrigues the youngest children, most convincingly told as a light story. Paperback.

MARKLE, SANDRA. *Exploring Summer.* New York: Avon, 1991. Directions for making a simple solar cooker and a sun tea recipe are given under "Cooking with Sunshine." Paperback.

MURPHY, PAT. *Bending Light.* Boston: Little, Brown, 1993. Find out from this engaging book how to make a magnifying lens with lemon jello! Try some of the other neat lens activities devised by the San Francisco Exploratorium Museum staff.[*]

ONTARIO SCIENCE CENTRE. *Having Fun with Magnifying.* Toronto: Kids Can Press, 1987. This activities book includes making and using a simple, failure-proof magnifier. Paperback.

PAUL, ANN. *Shadows Are About.* New York: Scholastic, 1992. This simple book brings shadows into young children's awareness. Paperback.

SIMON, SEYMOUR. *Shadow Magic.* New York: Lothrop, 1985. Children playing with shadows are led, through questioning, to discover basic information about shadows. Directions for making a shadow clock are included.

WYLER, ROSE. *Raindrops and Rainbows.* Englewood Cliffs, NJ: Simon & Schuster, 1989. Simple text and experiments explore aspects of separating light into rainbows.

Storytelling with Lights

1. Tell Leo Lionni's story, *Little Blue and Little Yellow,* using a small study lamp and blue and yellow plastic paddles to blend and separate the color families in the story. Let the children experiment with the light and paddles afterward. Have a sheet of white paper under the lamp. A set of colored plastic construction squares can also be used this way.

2. Use a small high-intensity lamp and three-dimensional objects to animate and tell a simple shadow story. Do this on the floor of a darkened room with the children sitting in a circle around the story setting. Small dolls can be children playing shadow tag outdoors. A cardboard tree or a wall of blocks can form a shady place for the dolls to rest after the game. A cardboard cloud could pass between the dolls and the lamp to perplex the story children. They can't find their shadows for the tag game and think they've lost them. Walk the dolls over to the children watching the story to ask them for help finding the shadows.

Poems (Resources in Appendix A)

There are many poems about light.

From *A Light In The Attic* by Shel Silverstein: "Shadow Race" and "Reflection"

From *Poems to Grow On,* compiled by Jean McKee Thompson:
About shadows: "Shadow Dance" by Ivy O. Eastwick
About shadows: "Kick a Little Stone" by Dorothy Aldis
About reflections: "Mirrors" by Mary McB. Green
About the spectrum: "I Wonder" by Virginia Gibbons

From *Poems Children Will Sit Still For,* compiled by Beatrice deRegniers:
About reflection: "A Coffeepot Face" by Aileen Fisher
About shadows: "8 a.m. Shadows" by Patricia Hubbell

From *A Child's Garden of Verses,* by Robert Louis Stevenson: "My Shadow" may need a bit of translating for children who know babysitters, but not nursemaids. It should be part of their poetry experience.

From *I Thought I'd Take My Rat to School,* compiled by Dorothy Kennedy, read Lillian Moore's "Recess," about children scribbling their shadows on the schoolyard.

Art Activities

A Reflecting Collage. Let children use pieces of foil gift wrap, aluminum foil, sequins, glitter, and gummed stars to make collages.

Reflector Hanging. Let the children cut pairs of simple shapes from foil paper or aluminum foil. Help them place three or four shapes face down in a line. Put a few drops of white glue in the center of each shape. Place a 15" (38 cm) length of string on the glue spots, then let the children cover each shape with its mate. Tie a gold plastic thread spool to the bottom of the string. Tape to the ceiling or to a door frame.

"Stained Glass" Medallions. Cut apart separate rings of plastic "six-pack" holders from frozen juice or drink cans. Pierce a hole at the top of each ring with a heavy threaded needle. Tie a thread loop. Let children dot glue around the ring and press colored tissue paper and cellophane over it; then trim away excess paper. Hang the medallions in the windows. Does light come through?

Waxed Paper Translucents. Let children cut shapes from colored tissue paper and arrange them as they wish between two sheets of waxed paper. Children seal the papers together by placing them on a newspaper-covered food warming tray and rubbing firmly across the waxed paper with a pizza roller.

Color Blending. After reading *Little Blue and Little Yellow* by Leo Lionni, children may be interested in blending paint colors. Younger children can try two primary colors at a time. Older children can try to create all the colors of the rainbow with three primary colors.

Dramatic Play

Ship Play. The crew of a building-block ship will enjoy using a periscope. Let them see how mirrors reflect light to create this interesting effect. Instructions for making a periscope are found in *Sun and Light,* by Neil Ardley (see Resources).

Shadow Fun. Draw the shades during active playtime on a rainy day and set up a projector at one end of the room. Turn on dancing music and let the children enjoy creating shadows as they dance in a designated area. Play shadow tag outdoors with the children on a sunny day.

Table Play

Reflection/Refraction Attraction. Follow up the reflection and refraction experiences by making as many of these fascinating toys and practical items available as you and your class can bring in to share. Kaleidoscopes, octascopes, and homemade milk-carton periscopes all depend upon mirrors within to create intriguing patterns and views. The Dragonfly Optiscope has 25 planes in its lens to imitate that insect's view of the world. Diffraction-grating lenses and novelty items produce rainbow effects. One popular miniature figure has such a lens that creates an appearing/disappearing face. These items are usually sold in nature stores and museum shops. Add old binoculars, varied plastic card magnifying lenses and other magnifying lenses, and a container of small things to enlarge and examine.

Creative Movement

Be My Mirror. Suggest to the children that they pretend to be your reflection in a mirror. They silently move as you move. You may be surprised at the number of movements you and the children can invent while you are seated, using your head, neck, shoulders, arms, hands, and fingers. This is a good activity to calm a restless group.

Creative Thinking

What If. "What if none of the lights could be turned on in your house one night? How could things be identified in the dark?" Pass a feeling bag containing several objects that can be identified with closed eyes by feeling or smelling.

SECONDARY INTEGRATION

Maintaining Concepts

1. Point out reflections when you notice them in the classroom: in doorknobs, in a child's shiny brass button, in the spoons at lunchtime, or in the rain puddles on the playground.

2. Hang a prism in a classroom window that gets direct sunlight. Experiment with balance and location to produce a good spectrum.

3. Use the opportunities that arise to comment on the need for light to see well: when raising the shades after rest time, when looking closely at a scraped knee, and so on. Comment that we need light to see our wonderful world.

4. Store magnifying glasses in a place where children will have access to them when they want to examine something. If you have a classroom microscope, let the children know that you will get it out for them when they want to examine a specimen. Talk about reflecting more light onto the specimen with the adjustable mirror.

5. Keep track of the length of the shadow made by a special tree or building near the school playground. Use a chalk mark on a paved surface, or stones on a grassy area, to compare its length when you are outdoors in the morning, at noon, and in the afternoon. Assign this responsibility to specific children, who then demonstrate the change to the others.

6. When the subject of lasers is brought up by children in terms of the destructive weapons they see in television programs or in certain toy advertisements, balance these ideas with the constructive uses of laser light. Tell them that this form of synthetic light is used by doctors to heal people. It is a powerful tool with many good uses. It is also used to "read" the special marks on grocery packages at the supermarket checkout counter they often pass. A tiny laser beam shines on compact discs to produce sound.

Connecting Concepts

Relate the bending of light beams to the surface tension experiences (see Chapter 8, What Is Water?). Recall that the pull on the outside of the water drop makes it curved. Light coming through this curve of water bends and changes the way in which we see through it. Surface tension keeps soap bubbles curved, and lets us see rainbow colors of spread-out light beams. In caring for classroom plants, remind children that all plants need light to stay alive and grow. Animals and humans need light to live, too, because they need plants for food.

Family Involvement

Encourage families to take their children to these features, if they are available in your community: an observatory, a lighthouse, stained glass windows, electric-eye door openers. Point out solar panels on home roofs that store the sun's energy for heating or convert solar energy to electricity. Families may have pocket calculators that use photovoltaic material to change light into electricity. Families can show children the laser scanner units at the supermarket and point out how the laser signals the price of groceries to the computer cash register.

RESOURCES

ARDLEY, NEIL. *Sun and Light*. London: Franklyn Watts, 1983.

BAINS, RAE. *Light*. Mahwah, NJ: Troll, 1985. Includes a brief note on laser light.

BRANLEY, FRANKLYN. *The Sun: Our Nearest Star*. New York: Harper & Row, 1988. Includes illustrations of solar panels using the sun's energy to heat a house. Paperback.

HANN, JUDITH. *How Science Works*. Pleasantville, NY: Reader's Digest, 1991. The carefully written chapter on Light and Sound presents clear, understandable background information on light as a form of energy, shadows, reflection, and refraction.

LEVENSON, ELAINE. *Teaching Children About Physical Science* (pp. 94–124). New York: TAB Books, 1994. Background information for teachers is set out clearly in brief definitions of terms.

MACAULAY, DAVID. *The Way Things Work*. Boston: Houghton Mifflin, 1988.

VAN CLEAVE, JANICE. *Microscopes and Magnifying Lenses*. New York: Wiley, 1993. This book of science fair projects for advanced students provides good background information about lenses.

Teaching Resource

SUNPRINT KIT:

Create images using sunlight and light-sensitive paper. Museum stores or nature stores carry these. Also may be ordered from Lawrence Hall of Science (see Appendix A).

16

The Environment

Plants, animals, air, water, weather, rocks, and the human body are the primary elements of the environment that are accessible to children. Learnings from these topics can serve as the basis for constructing an understanding of the environment, the system in which each of us participates. The experiences suggested for this chapter will help children create connections between concepts studied earlier. To help develop an understanding of the overarching concept of the environment, these subconcepts will be explored:

- There is an interconnectedness among things: plants, animals, air, water, weather, rocks, and ourselves.
- The environment is where we are. We can study it, as well as live in it.
- We can work together to sustain the environment by restoring, reusing, repairing, and recycling.

These concepts can be presented after the primary elements have been studied, or they can be used as an introduction to the year's science study, since they provide a framework for considering the other elements. Because the broad topic of the environment is complex, it is better suited to second and third graders than to younger children. Some of the activities can be successfully used with younger children. Teachers can best decide for their own classes when and how to use them.

In the experiences that follow, children will explore connections among themselves and their surroundings, examine a microclimate carefully, construct terrariums, develop an affinity for a tree, and participate in recycling in various ways.

CONCEPT: **There is an interconnectedness among things: plants, animals, air, water, weather, rocks, and ourselves.**

1. What is connected to what? (Indoor activity)

LEARNING OBJECTIVE: To develop an awareness of interconnectedness among things.

MATERIALS:

Index cards

Paper clips or tape

Markers

Scissors

Ball of string or yarn with some 6' (about 6m) lengths precut

Bulletin board or wall space as background for web (or form web on the floor within a circle of children)

SMALL-GROUP ACTIVITY:

1. Ask children to tell what they know about the word *connections*. You may need to offer some examples, such as LEGO blocks or train cars being connected to one another, or give one end of a string you are holding to a child: "Now you and I are connected." Ask children to think of connections they have to other persons in the class, or in their families. Use the connections they offer to begin shaping a web of index cards and string. For instance, if Sarah says she is connected to her mommy, write *mommy* on one card, *Sarah* on another card. Tape the cards to the background and attach a length of string between the cards. Perhaps Sarah's mom drives a car pool for Tyler, so a Tyler card is taped to the background. Mom's card is connected to Tyler's card with another length of string, and so on. If Tyler goes to play at Andrew's house after school, they are connected. Continue to join strings and cards as long as you can keep it up. Answers will vary, but a web of connections should develop from the conversation.
2. To ensure that each child has a connection of some kind firmly in mind, accept responses for all kinds of connections. Discuss things outdoors that children can think of as connected.
3. Leave the web in place for children to study. Leave extra string and cards nearby for children to add new ideas. Older children can also add written explanations of the connections they make.

2. *What is connected to what?* (Outdoor activity)

LEARNING OBJECTIVE: To develop an awareness of the interconnectedness of the outdoor environment.

MATERIALS:

Index cards or sticky labels, 3 or 4 per child

Pencils

Sheets of cardboard or heavy fabric or plastic for background, one per group of 3–4 children

SMALL-GROUP ACTIVITY:

1. Indoors, recall the first "connections" experience with children. "Today we are going outside to study the connections between things around us in our schoolyard *environment*. Where we are and everything that surrounds us here are called an environment." For primary children, form into cooperative learning groups. For preschool–K children, do both steps in small groups. Have children collect supplies.
2. Outdoors, the children sit in a circle and brainstorm about things they can see, hear, smell, and touch, as well as other things they know exist in this particular environment (the air, a nearby creek, etc.).

3. Work with younger children to make cards for the things children have discussed. Join them into webs of relationships, attaching the webs to the background material. Older children break into their small groups to do this independently.

4. In a discussion, consider ways that webs of each group differ. Are there connections no one has thought of yet? Are some connections more direct than others? One purpose of the discussion is to raise questions that individual children may want to pursue. "What if . . ." questions can provoke fresh thought. "What if there were no water? What if the sun shone all day?" Encourage independent questioning. Follow up on individual children's queries (e.g., "What do ants eat?") to help them find classroom books and other resources.

3. Living in the web of life.

MATERIALS:

Ball of string

LARGE-GROUP ACTIVITY:

Have children stand in a circle. You stand in the middle with the ball of string in hand. Ask a child to name a plant growing in the area. Give that child the end of the string to hold. Ask: "Who eats that plant?" Take the string to a child whose response is reasonable (e.g., rabbit), so the rabbit can be connected with the string to the plant. Continue with a food or shelter question (e.g., "What eats the rabbit?" "Where does the fox live?" etc.) until everyone is connected in the web. Ask a child (perhaps one who mentioned a tree) to sit down. Ask: "If the tree falls down, who feels the tug on their string?" etc. Through this activity, children can physically experience interconnectedness.

Read Alvin Tressalt's book *The Gift of the Tree,* or Natalia Romanova's book *Once There Was a Tree.*

CONCEPT: **The environment is where we are. We can study it, as well as live in it.**

1. What is the environment of creatures like?

MATERIALS:

String or yarn cut to 4' (or 4 m) long, one per child

Magnifiers, one per child

SMALL-GROUP ACTIVITY:

1. "Do you know the story about Thumbelina? She was as tiny as your thumb. Let's see what it would be like to take a Thumbelina walk." Take children to a nearby grassy or wooded area. Let each child

Where do you think this little guy lives?

choose a section to explore. "Lay out your string and crawl along next to it, pretending you are a tiny creature. Use your magnifying glass, and keep your head down low to see every detail."

2. "Who do you see in this little world? Are there other small creatures? Are they friendly? Is there a drop of water to drink, or anything to eat? Where would you like to rest here?"

3. Children can report to the group what they saw. Encourage children to draw their micro-worlds afterward.

Read about a child lying in the grass exploring the eye-level world nearby in Rose Wyler's book, *Grass and Grasshoppers*. Read about a place "where clover tops are trees" in Robert Lewis Stevenson's classic poem, "The Little Land" from *A Child's Garden of Verses*. (See Appendix A.)

2. Can we make a micro-climate?

LEARNING OBJECTIVE: To construct and sustain a terrarium.

MATERIALS:

Tiny plants dug from the grassy area, if permitted; or small seedlings started earlier as class activity (see p. 56)

Bits of moss

SMALL-GROUP ACTIVITY:

1. "Can you imagine making a tiny world where it rains by itself? Plants can live there, and you could pretend to live under a tiny leaf." Show children how to just cover the bottle base with a layer of pebbles, cover them with bits of crushed charcoal, and then spoon in a deeper layer of soil.

Small sticks or bark bits, shells, small rocks

2-liter soda bottle with detachable base for each pair of children, or each child

Potting soil

Gravel

Crushed charcoal bits (*not* chemically treated briquettes)

Old spoons

Newspapers, if working indoors

Spray bottle of water

GETTING READY:

Prepare bottles: Soak off labels with warm water. To remove base, soften glue with hot water; or warm on radiator or microwave briefly; or place in freezer to release glue.

Use table knife to pry off base.

Use scissors or blade knife to cut off top 1/3 of bottle. (Recycle tops as funnels.)

Carefully arrange a few small plants in the soil: First make a hole for each rooted plant. Then press soil down over roots. Decorate with bark or other natural objects. A shell could be a tiny pond.

2. Dampen plants with a few sprays of water. Invert the bottle over the garden, tucking the sides into the base to form a dome. Place in a warm, light place, but not in direct sunlight.

3. The next day watch for evidence of water evaporating and condensing inside as drops of "rain" that return to the soil. (See chapter 9, p. 168.) If children sprayed their plantings too generously, the domes will be covered heavily with condensation. Lift off those dome tops for several hours. If no moisture condenses inside the dome, spray plants again. Talk about the self-sufficiency of this micro-world. Think about how these plants get their food and water.

3. What can we learn about a tree without using our eyes?

LEARNING OBJECTIVE: To develop an affinity for a tree; to be able to see trees as individual organisms.

MATERIALS:

Freestanding trees (without poison ivy growing around them)

Blindfolds for half the group

SMALL-GROUP ACTIVITY:

1. Outside with the children ask: "Would you like to get to know something old and big? It's something that can't talk or see, so we'll get to know it without looking at it or talking to it. Have any ideas what this thing might be?"

2. Have children work in pairs. One child leads the blindfolded partner to a tree. The blindfolded children feel the tree with their hands. Encourage multisensory exploration. Ask: "What is the bark like? Do you find branches? Are there leaves? Are roots sticking out? Are there holes in the tree? Can you reach all the way around the tree trunk? How does your tree smell? Does it feel sticky?"

3. When the blindfolded child is through, the partner leads her or him away from the tree, rotates the child a bit, and removes the blindfold. The child can try to recognize his or her tree. Reverse the roles of the children.

4. Indoors, let children draw how they remember their tree. Encourage them to keep track of their

FIGURE 16–1
Bottle terrarium

special trees through the school year. Children become attached to individual trees in this way, and begin to think of the woods as alive, not merely backgrounds to life. Discuss the relationships of trees to other animals and to ourselves (see Connecting Concepts, p. 99).

Read Joanne Ryder's book *Hello Tree.* (See Children's Books and Resources.)

Read Harry Behm's poem "Trees" from *Land, Sea and Sky* by Catherine Paladino.

CONCEPT: **We can work together to sustain the environment by restoring, reusing, repairing, and recycling.**

LEARNING OBJECTIVE: To become aware that we can reduce our consumption of our world's resources.

MATERIALS:

Discarded cardboard cartons

Markers

Bathroom scale

LARGE-GROUP ACTIVITY:

Part 1: Reducing Waste

1. Ask children to recall all of the things they throw into the wastebasket at school. List the items on the chalkboard. Ask children to group their responses. A variety of groups is possible. Provide and label a carton for each category, so children can collect separately each of those throwaways for several days. (Alert the custodial staff not to empty the boxes at day's end.)

2. At the end of the collecting period, weigh and graph the amounts of trash in each carton. Extend the graph on the chalkboard to show what the amounts would look like after 9 months of school. (You provide the calculations.)

3. See if children think there is a problem. Have groups of children brainstorm how to reduce the amount, and reuse different types of trash. If they do not think that there is a problem with trash, ask: "What would happen if the garbage truck never came to our school to whisk our trash out of sight? What did people do with trash in the days before there were garbage trucks?" Consider ways to reduce waste in the classroom, such as cutting paper towels in half, keeping a box of paper used on one side for reuse by children, cherishing pencils and crayons. Plan with children how to carry out their ideas on reducing waste. Almost every idea is worth trying for a day or two. Some ideas will really work. Keep a running record, or estimate how much waste was avoided by the class at the close of the school year. Consult *50 Simple Things Kids Can Do to Recycle* for other ideas. (See Resources.)

4. Undertake a school building and yard environmental improvement activity. Figure out together what might make this environment a cleaner, more life-enhancing place to be. Talk about everyone's part in taking care of our schoolyard environment by "putting waste in its place." Think of a slogan for the school to adopt to help remember that "every litter bit hurts." Such environmental activism can be schoolwide, providing interaction between younger and older students for their mutual benefit. Read *Miss Rumphius*, by Barbara Cooney, as an example of how one person can enhance the environment. Many state governments provide resources for habitat improvement, particularly Fish and Wildlife departments. (See Resources.)

5. Explore the concepts of biodegradable and non-biodegradable in terms of materials that can or cannot easily rot to become part of the soil. In a place on the school grounds that won't be disturbed, bury several items that won't degrade, such as a styrofoam cup, a plastic bag, and a soda can, along with a paper bag. Check several months later to see how different things were affected by the damp soil. Encourage children to make the connection between what they observed and our need to recycle and reuse man-made materials that won't degrade. Start a compost pile in the schoolyard. (See Soil Composition Relationships, p. 72.) Read the last half of Vicki Cobb's book, *Lots of Rot,* to amplify the discussion.

6. Put broken classroom equipment in a fix-it box to be repaired instead of discarding such items. Children gain satisfaction and a sense of importance from making repairs to restore equipment. Prevail upon the custodian, or an amenable parent, to restore things that are beyond the children's or your own skills. Older children can tape torn book pages. They can learn to do simple sewing repairs, if you have an emergency kit with safety pins, thread, and large enough needles. Preschoolers and kindergarteners take pride in maintaing school bicycles. Keep an oil can handy for squeaky axles, and a crescent wrench for tightening handlebar bolts.

Part 2: Conserving water

Introduction: Talk about the crucial role of water in ecological relationships. It is one of the nonliving substances that all living things depend on to stay alive. It is used over and over again in nature. Recall the evaporation, condensation rain cycle in the terrarium domes. (See Evaporation and Condensation Cause Precipitation, p. 168.) When too many waste materials are emptied into lakes, streams, and rivers, or when chemicals seep into underground water supplies, living things may have less water than they need. *No way has ever been discovered to make new water,* so we must use the world's supply carefully to keep it clean. Investigate the way water is cleaned in your area.

1. Try out a simple way your children can keep from wasting this resource. Using empty half-gallon milk containers, collect the amount of water that runs from a school sink for 1 minute. Translate this amount in terms of numbers of glasses of water that can be wasted each time children wash their hands under running water.

Afterward, talk about using a stoppered basin of water for handwashing. Talk about the need every person, animal, plant, tree, and the land itself, has for water. Children who do *not* let water run needlessly are helping our planet Earth.
2. As aware stewards of Earth's non-renewable resources, celebrate Earth Day, April 22!

PRIMARY INTEGRATING ACTIVITIES

Math Experiences

1. Use found natural objects for ordering by size: twigs, stones, leaves, feathers. Interesting questions can arise when working with natural objects that manufactured manipulatives do not present. For instance, is a short thick stick "smaller" than a long, thin one?

2. Use found objects to count big numbers. How many autumn leaves can you fit into a berry basket? How many does it take, stacked one on top of the other, to make an inch-high stack? How many leaves are on a tree? How many acorns? How many seeds in a pod? Point out the abundance of nature. Read *How Much Is A Million?* by David Schwartz.

3. Compare leaf areas. Give children sets of various sizes of cardboard and ask them to find leaves that approximately match each piece. Older children can work with graph paper to estimate how many square units each leaf covers.

4. Estimate measurements. Count out the steps it takes to walk around the perimeter of the schoolyard or blacktop area. Does everyone get the same measurement? Discuss differences. How could those measurements be standardized?

5. Look for symmetry in natural objects. For instance, are both sides of a leaf alike? Look at deciduous trees after the leaves have fallen. Are they symmetrical? Compare free-standing trees with those in a more crowded area.

Music (Resources in Appendix A)

Listen to the Tickle Tune Typhoon group sing "A Place in the Choir" on their *Hug the Earth* cassette. It cheerfully affirms every being's right to live. "Oh Cedar Tree" is a Native American chant that even the youngest children could sing well.

Listen to several ecology songs in Mary Miche's *Nature Nuts* cassette, including "Bats Eat Bugs," "Recycle Blues," and "Garbage."

Raffi's cassette, *Raffi on Broadway,* has numerous songs about the Earth, the most singable being, "Big, Beautiful Planet," "May There Always Be Sunshine," "One Light, One Sun," and "KSE Promise Song," the song for a children's environmental group.

Stories and Resources for Children

AYRES, PAM. *When Dad Fills in the Garden Pond*. New York: Alfred A. Knopf, 1989. The realistic story of a family considering the value of a messy, fun pond against that of a pretty flower garden.

Brother Eagle, Sister Sky. Attributed to Chief Seattle. (See p. 285, Project Learning Tree) New York: Scholastic, 1992. Wonderful illustrations by Susan Jeffers accompany this beloved environmental credo.

BURTON, JANE. *Animals Eating*. Brookfield, CT: Newington Press, 1991. Together with the other books in this series—*Animals Fighting* and *Animals Talking*—shows processes shared by people and other animals. These books help children see the relatedness of all life.

CADUTO, MICHAEL J., & BRUCHAC, JOSEPH. *Keepers of the Earth: Native American Stories and Wildlife Activities for Children*. Golden, CO: Fulcrum Publishers, 1991. The environmental emphasis is well-illustrated.

CHERRY, LYNNE. *The Great Kapok Tree*. San Diego: Harcourt Brace Jovanovich, 1990. A woodsman is persuaded not to destroy the rich natural community of the great kapok tree in the rain forest.

CHERRY, LYNNE. *A River Ran Wild: An Environmental History of the Nashua River*. San Diego: Harcourt Brace Jovanovich, 1992. A beautifully illustrated story of a river that is pristine, becomes polluted, and is restored.

COBB, VICKI. *Lots of Rot*. New York: Lippencott, 1981. Read aloud selectively to develop appreciation for the role of decomposers in the soil renewal cycle.

CONE, MOLLY. *Come Back, Salmon*. San Francisco: Sierra Club, 1992. The inspiring true story of an elementary school in Washington State that cleaned up a creek and restored the salmon migration.

COONEY, BARBARA. *Miss Rumphius*. New York: Viking, 1982. This book tells how one person made her environment more beautiful by planting flowers in a nearby meadow.

GIBBONS, GAIL. *Recycle! A Handbook for Kids*. Boston: Little, Brown, 1992. A sprightly introduction to the topic of recycling, with basic suggestions for action at the end.

GILMAN, PHOEBE. *Something from Nothing*. New York: Scholastic, 1992. A resourceful grandfather and boy keep reusing his blanket until nothing remains (but check the mice in the border illustrations).

HUGHES, SHIRLEY. *The Big Alfie Out of Doors Story Book*. New York: Lothrop, Lee, & Shepard, 1992. Alfie and his family enjoy being outside: camping, playing store, walking in the country, cherishing a special rock, playing on the beach.

KITCHEN, BERT. *And So They Build*. Cambridge, MA: Candlewick Press, 1993. A lovely text and striking, detailed paintings provide a respectful look at the intricate shelters certain animals build.

MAZER, ANNE. *The Salamander Room*. New York: Alfred A. Knopf, 1991. In this charmingly illustrated book, a conversation between a child and his mother imaginatively informs children about the shelter and food requirements of the salamander.

ROMANOVA, NATALIA. *Once There Was a Tree*. New York: Dial, 1989. This lovely ecosystem tale from Russia is enriched by soft, detailed illustrations. New York Times Notable Books of the Year Award. Paperback.

SCHWARTZ, DAVID M. *How Much Is a Million?* New York: Lothrop, Lee, & Shepard, 1993.

TRESSALT, ALVIN. *The Gift of the Tree*. New York: Lothrop, Lee, & Shepard, 1992. A fine story about the hundred-year life cycle of an oak tree as it shelters and promotes new life.

WYLER, ROSE. *Grass and Grasshoppers*. New York: Julian Messner/Simon & Schuster, 1990. The text and illustrations move from good feelings about lying in grassy fields to wondering, observing, experimenting, and learning about grass and the small creatures it shelters.

YOLEN, JANE. *Letting Swift River Go*. Boston: Little, Brown, 1992. An account of an actual event: the damming of a river to provide water for Boston and its effect on the people who lived by the river.

Poems (Resources in Appendix A)

From Robert Lewis Stevenson's *A Child's Garden of Verses*, read "The Little Land."

From Catherine Paladino's *Land, Sea and Sky*, read "Trees," by Harry Behm.

Art Activities

Modeling Beeswax. Children soften pieces of beeswax by warming it with their hands until it is pliable. They can model and shape it as desired, stretching it thin for translucency. Beeswax hardens when it cools. It can be reused by softening it again. Because it melts, avoid putting finished work in direct sunlight, or leaving it in a hot car or other places. This natural material is obtainable in colors from craft stores. For an ecology experience, buy a piece of natural comb honey and let the honey drip out of the comb. What remains is beeswax, in a very sticky state. Wash it carefully with warm water and use it for modeling.

Found Clay Making. Try to locate a natural source of clay; a streambank, a ditch, or an eroded place are often good sources. (Ask a local potter for suggestions.) Dig out some clay together with the children, if possible. Notice the texture. Are there rocks in it? Do the children have ideas about how to make it as smooth as commercially available clay? One way is to soak the clay in a bucket of water for several days, until it becomes liquid. Strain it, and allow it to settle into layers. This process takes several days, so be patient. When the clay dries somewhat, it can be worked and shaped.

Sun Prints. Arrange leaves or other nature finds on dark-colored construction paper. Weight it down with rocks on the leaves and on each corner of the paper. Leave for about an hour in bright sunlight. The paper shaded by the objects will stay the same color, while the rest of the paper will be faded by the sunlight. Solar print (light sensitive) paper is commercially available for the same purpose. Follow the package instructions.

Outdoor Chalk Making. Mix powdered tempera paint with 6 tablespoons of plaster of paris. Stir in 1 cup of water. Pour quickly into a plastic ice-cube tray and let it harden. This will take about 1/2 hour. Children can use the chalk on the sidewalk or asphalt.

Dramatic Play

Housekeeping Play. If your environment permits, let children set up "houses" in the nature area. Branches, leaves, and stones are the ingredients for imaginative house play that encourage children to value natural objects. If you can, trim bushes so that children can use them for refuge and privacy. (Drooping branches of a large forsythia bush work well for this.) When children interact with nature in a way that is meaningful to them, they can develop the attitudes that later lead to responsible stewardship. Read "Shop" in *The Big Alfie Out of Doors Story Book*, by Shirley Hughes, to inspire new play ideas. Alfie and Annie Rose set up a store, selling leaves, sticks, and such.

Creative Thinking

Classifying. Take children outdoors. Give them paper bags and instructions to work in pairs to gather up whatever looks like trash. (Leaves growing on bushes are not trash.) Back in the classroom, provide trays for children to sort their found objects into categories. Accept all categories and the reasoning behind them. Later, ask children to re-sort things into two groups: things people made, and things nature made. Discuss what would happen to these objects if they were left outside. Send recyclable man-made things to a local recycling facility. Add biodegradable things to your compost heap.

Inferring. Play the "Guess My Name" game. Obtain small pictures of animals that are common to your area: small mammals, birds, insects. Put them in a bag. Show a small group of children how to play the game: Choose a picture from the bag, but don't look at it. Ask children questions for clues about what it might be: "Am I gray?" "Do I live in a tree?" "Do I eat seeds?" Infer from children's answers the name of the mystery animal. When children understand how to play, the pictures can be taped onto their backs, so that several children can play simultaneously. The hard thing, of course, is not to tell the name right off!

Food Experiences

Having children gather and prepare food that nature provides can be an appealing and challenging prospect for nature-loving teachers. In our regions, lucky teachers will find mulberry trees nearby to provide an edible wild treat in early summer. Some may know where an accessible wild grape vine or elderberry bush has ripened fruit ready to harvest in fall. Both of these fruits are too tart to eat freshly picked, but an ambitious teacher can gather them (both are out of reach of young children) and make jelly to share with the class. Photographs of the teacher at work could authenticate the harvesting from the wild state.

Children love to gather fallen nuts and attractive berries. They do so indiscriminantly, often sampling their treasures. When a puffball unexpectedly erupted in our school playground, it seemed repulsive to the children, though it is edible and prized as a delicacy. Conversely, they were attracted to the bright red nightshade berries and lush blue pokeberries that were clearly visible beyond the playground fence. Both are poisonous. **As a safety precaution, always check with a local naturalist before offering any food from the wild to children.** (See the note on p. 64 regarding common poisonous plant materials.) A reliable resource for learning more about foods in the wild is *Stalking the Wild Asparagus* by Euell Gibbons. (See Resources.)

For the longer range, see Robin C. Moore's *Plants for Play: A Plant Selection Guide for Children's Outdoor Environments.* He offers many ideas for landscaping the schoolyard with edibles.

Tell children that many early Native Americans relied only on wild fruits and plant foods to add to their food supplies, since few groups cultivated crops. Later settlers also made use of abundant wild fruits, since it was more important to grow grains and other kinds of crops at first. Grocery stores heaped with cultivated fruits and vegetables came into being much later in our history.

Field Trips

Creek Walk. If there is a creek nearby to walk along for a way, make the trip by yourself in advance to be sure it is suitable for your group. Take plenty of time to notice things that can be seen along the way: plants, insects, rocks, clay out-croppings, trees. Creeks tend to be low in the fall, which makes them safer for young children. As we have said earlier, keep the science focus simple. There will be too many exciting things to see to make lengthy explanations bearable for the children. Be sure that the adults who accompany the class trip understand this. Children's previous constructed knowledge of elements of the environment may come to light here, and in classroom discussions afterward.

Planetarium. Older children are often deeply interested in space, and some would enjoy a planetarium visit. In this context they can see evidence that our planet, big as it is to us, is a tiny part of a vast system. This is both provocative and reassuring: "We are part of a larger design, and will be safe."

Ecological Scavenger Hunt. Margaret Drysdale takes her classes to an empty lot where they check the area for debris. They collect these materials and classify them into two piles: man-made materials and natural materials. Ms. Drysdale explains that only the things from nature can be left on the ground, be-cause they will decompose and eventually become part of the soil. The children carry the man-made materials back to the school recycling/trash bins, because those things won't decompose. She emphasizes that a discarded aluminum can or plastic item could still be littering the land when the children are old enough to be grandparents.

Landfill. Some children might be interested in seeing where solid waste is contained. The utter blandness and extensiveness of the area can facilitate discus-sion of recycling and reusing. Follow-up play in the sand table can be facilitated with toy trucks and things to bury. Since landfill visits are restricted to viewing from a distance, this experience would have to be one suggested to families. Children who have been able to see the landfill would then report back to the group what they observed. Joanna Foster's book, *Cartons, Cans, and Orange Peels: Where Does Your Garbage Go,* has good photographs and information to prepare children for this trip.

Sewage Treatment or Water Treatment Plant. Some children would like to know what happens to the dirty water after washing their hands or flushing the toilet. Ask for children's ideas about satisfactory ways of dealing with dirty water and ways to make water from lakes, rivers, or underground sources clean and safe enough for us to drink. It may be possible in your area to arrange a visit to the treat-ment plants. Make the process more concrete by exploring the concepts of filtering and sedimentation. Collect a jarful of muddy water from a puddle, or create some by adding a scoop of playground soil—gravel and other flotsam included—to a quart of water. Line a funnel with a coffee filter and pour some of the muddy wa-ter through it into a smaller container. Check the results. Identify the filtered water as *cleaner,* but *not drinkable.* Let the remaining water stand for a day in the quart jar to observe how the dirt and other stuff settles to the bottom as sediment, with the water above clearer and cleaner.

Read aloud *The Magic Schoolbus at the Waterworks,* by Joanna Cole.

SECONDARY INTEGRATION

Maintaining Concepts

1. Have children work in cooperative learning groups, structured so that children are truly interdependent. Remind them that they are like the environment, where every part is connected, and each part depends on other parts being there and functioning.

2. Assign different children each day to care for classroom pets and plants, maintaining the necessary micro-environments.

3. As children begin to appreciate trees as living things and begin to understand that trees play a part in the oxygen cycle, offer an important statistic from Project Learning Tree: "A 12-year-old child needs to plant and maintain 65 trees in order to offset the amount of carbon dioxide that child will put in the atmosphere during the rest of his or her lifetime." (See Resources, p. 286.) Try to arrange to plant a tree seedling for the class, and plan for watering and otherwise maintaining the new trees. Our schools have done this for many years, so today's youngsters have shadier, screened-off playgrounds.

Connecting Concepts

Generally speaking, the environment encompasses every element, and everything that exists in and around our planet. It would be easy to overwhelm youngsters by trying to pull together a multitude of possible connections in classroom discussions. For this reason, concept connections in this chapter have been limited to concepts found in Chapters 4 through 10. The topics of endangered species of animals, large-scale pollution, and depletion of nonrenewable resources have been avoided deliberately. It is our belief that issues of such scope needlessly burden and sadden young children who are helpless to remedy them. The concepts in this chapter focus on the positive connections in nature, since children and adults are more likely to take care of what they truly appreciate.

Family Involvement

There are many things that families can do to make children more sensitive to the environment. Probably the most important of all is for adult family members to model such values. They can involve children in recycling programs going on in their area. Children can participate by washing empty cans and stacking newspapers. Buying bulk foods at the store reduces the amount of material needed for packaging. Children can help with the measuring and weighing to gain experience in valuable math processes. Families could point out the community landfill site, if it can be seen from the car. Children can participate in energy conservation by turning out unneccesary lights, not letting faucets run needlessly, and walking or biking instead of expecting to be driven short distances. Many communities have stream, roadway, or beach cleanup programs. Newsletters to families can announce the dates and locations.

RESOURCES

BAKER, ANN, & JOHNNY. *Counting on a Small Planet: Activities for Environmental Mathematics.* Portsmouth, NH: Heinemann, 1991. The "whole-math" approach is explained and ideas are suggested for getting children to use math naturally in exploring their environment.

CADUTO, MICHAEL. *Pond and Brook: A Guide to Nature Study in Freshwater Environments.* Englewood Cliffs, NJ: Prentice Hall, 1985. Many creative ideas for involving children with nature.

CLAYCOMB, PATTY. *Love the Earth: Exploring Environmental Activities for Young Children.* Livonia, MI: Partner Press, 1991. Many creative ideas for involving children with nature.

COLE, JOANNA. *The Magic Schoolbus at the Waterworks.* New York: Scholastic, 1993. Ms. Frizzle takes the class on her usual zany-style, informative trip through the waterworks.

EARTHWORKS GROUP. *50 Simple Things Kids Can Do to Recycle.* Berkeley, CA: EarthWorks Press, 1994.

EARTHWORKS GROUP. *50 Simple Things Kids Can Do to Save the Earth.* Kansas City: Andrews and McMeel, 1990.

FOSTER, JOANNA, *Cartons, Cans, and Orange Peels: Where Does Your Garbage Go?* New York: Houghton Mifflin, 1991. Various methods for handling garbage are described.

FRY-MILLER, KATHLEEN, & MYERS-WALL, JUDITH. *Young Peacemakers Project Book.* Elgin, IL: Brethren Press, 1988. Links stewardship with peace.

GIBBONS, EUELL. *Stalking the Wild Asparagus.* Brattleboro, VT: Allen C. Hood, 1988.

HERMAN, MARINA, SCHIMPF, ANN, PASSINEAU, JOSEPH, & TREUER, PAUL. *Teaching Kids to Love the Earth.* Duluth, MN: Pfeifer-Hamilton, 1991. A thoughtful book, replete with resources.

KARMOZYN, PATRICIA, SCALISE, BARBARA, & TROSTLE, SUSAN. 1993. *A Better Earth: Let It Begin with Me!,* Childhood Education, 69, 225–230.

KOHL, MARYANN, & GAINER, CINDY. *Good Earth Art: Environmental Art for Kids.* Bellingham, WA: Bright Ring Publishing, 1991. Many recycling ideas and homemade materials.

MCVEY, VICKI. *The Sierra Club Kid's Guide to Planet Care and Repair.* Good information for teachers, such as a child in Kenya may use 1 gallon of water a day to the U.S. child's 250 gallons.

MOORE, ROBIN. *Plants for Play: A Plant Selection Guide for Children's Outdoor Environments.* Berkeley, CA: MIG Communications, 1993. Many ideas for landscaping the schoolyard to provide children with edibles and play materials.

PROJECT LEARNING TREE. *Environmental Education Activities, K–6 Grade.* Washington, DC: American Forest Foundation, 1993. The origin of the beloved environmental credo attributed to Chief Seattle is examined on pages 344–347.

RUSSELL, HELEN ROSS. *Ten Minute Field Trips* (2nd ed.). Washington, DC: National Science Teachers Association, 1990. This book emphasizes using the urban school grounds for environmental science. The chapter on interdependence of living things is excellent.

Additional Resources

Environmental Education Activities for Grades K–6
> Project Learning Tree
> American Forest Foundation
> Washington, DC

Wild School Sites: A Guide to Preparing for Habitat Improvement Projects on School Grounds
Available free in every state from the Fish and Wildlife Departments
Or contact:
> Project Wild
> 5430 Grosvenor Lane
> Bethesda, MD 20814

Resources for Music, Recordings, Poetry, Creative Movement, and Equipment Ordering

MUSIC

Cassettes

BANANA SLUG STRING BAND, *Adventures on the Air Cycle*. Music for Little People (cassette). "Air Cycle Swing," "Lizard," "No Bones Within," Animals," "Ecology."

BERMAN, MARCIA. *Rabbits Dance: Marcia Berman Sings Malvina Reynolds*. B/B Records, 1985. 570 N. Arden Blvd., Los Angeles, CA 90004.

BERMAN, MARCIA, AND ZEITLIN, PATTY. *Spin Spider, Spin: Songs for a Greater Appreciation of Nature*. Educational Activities. Includes songs about lizards, birds, snakes, frogs, and insects.

CHENILLE SISTERS. *1 2 3 Kids*. Red House Records. Includes "The Kitchen Percussion Song": Making music with things found in the kitchen.

CROW, DAN. *A Friend, a Laugh, a Walk in the Woods*. Includes "Walking On My Wheels" in which a child uses a wheelchair to get around, "The Zucchini Song," "Blowing Up Balloons," "The Shape of My Shadow." Sony Kids Music.

MICHE, MARY. *A Kid's Eye View of the Environment*. Star Trek. Includes: "Spiders and Snakes," "You Can't Make a Turtle Come Out," "Bug Bits," "Dirt Made My Lunch."

MICHE, MARY. *Kid's Stuff*, Star Trek. Includes "The Cat Came Back."

MICHE, MARY. *Nature Nuts*. Star Trek. "Recycle Blues," "Pollution," "Hey Ms. Spider."

PALMER, HAP. *Walter the Waltzing Worm*. Educational Activities.

RAFFI. *Evergreen, Everblue*. Shoreline/MCA.

RAFFI. *Raffi in Concert*. Shoreline/MCA.

ROGERS, FRED. *Won't You Be My Neighbor?* Family Communications, Inc. audiocassette MRV 8106C.

ROGERS, SALLY. *Piggyback, Planet: Songs for the Whole Earth*. Round River Records.

SEEGER, PETE. *Birds, Beasts, Bugs and Little Fishes*. Smithsonian Folkways.

ZAMFIR, GHEORGE. *Romance of the Pan Flute*. Cassette #32150. Polygram Classics: 810 Seventh Ave., New York, NY.

Music Books

CARLE, ERIC. *Today Is Monday*. New York: Putnam & Grosset, 1993. Splashy illustrations capture the fun of singing this simple song about favorite foods.

HAINES, JOAN, & GERBER, LINDA. *Leading Young Children to Music* (4th Ed). Englewood Cliffs, NJ: Merrill/Prentice Hall, 1992.

RAFFI. *Everything Grows*. New York: Crown, 1989.

RAFFI. *Singable Songbook*. New York: Crown, 1987.

READERS' DIGEST. *Oats, Peas, Beans.* Pleasantville, NY: Readers' Digest, 1985.
REY, MARY. *Over in the Meadow.* New York: Penguin, 1986. Paperback.
SILBER, IRWIN. *Folksong Festival.* New York: Scholastic Book Services, 1974.

Resources for Ordering Cassettes

Chinaberry Books
2780 Via Orange Way, Suite B
Spring Valley, CA 91978
Tel: (800) 776-2242

Music for Little People
P.O. Box 1460
Redway, CA 95560
Tel: (800) 346-4445

POETRY

DEREGNIERS, BEATRICE SCHENCK, MOORE, EVA, & WHITE, MARY M. *Poems Children Will Sit Still For.* New York: Scholastic, 1973.
FISHER, AILEEN. *When it Comes to Bugs.* New York: Harper & Row, 1986.
FLEISCHMAN, PAUL. *Joyful Noise: Poems for Two Voices.* New York: Harper & Row, 1988. Awarded the Newberry Medal, this fine collection of poems captures the movement, sound, and essence of selected insects. Exquisite drawings by Eric Beddows reflect the imagery evoked by the poems.
HOPKINS, LEE BENNETT (Ed.). *Weather.* New York: Harper Collins, 1994. An I-Can-Read book. Brightly illustrated, carefully chosen, enjoyable poems about the weather.
HUGHES, SHIRLEY. *Out and About.* New York: Lothrop, Lee, & Shepard, 1988. Warm, joyful depictions of a child taking pleasure in the outdoor offerings of each season's weather. Tenderly illustrated in rich detail by the author. It includes poems about water-bouyancy, and a rainbow seen through a hose spray.
KENNEDY, DOROTHY (Ed.). *I Thought I'd Take My Rat to School: Poems for September to June.* Boston: Little, Brown, 1993. Humorous poems about school realities include the topics of light, animals, plants, and nutrition.
MILNE, A. A. *Now We Are Six.* New York: E. P. Dutton, 1961.
MOON, PAT. *Earth Lines.* New York: Greenwillow, 1991. "The Bird's Nest": children try to compete with a bird to build the best nest. The bird wins!
MOSS, JEFFERY. *The Butterfly Jar.* New York: Bantam Books, 1989. Includes "The First Muscician" and "The Banana King."
PALADINO, CATHERINE. (Ed. and illus.). *Land, Sea and Sky.* Boston: Little, Brown, 1994. Fine color photographs of nature accompany the well-chosen poems.
RYDER, JOANNE. *Inside Turtle's Shell, and Other Poems of the Field.* Englewood Cliffs, NJ: Prentice Hall, 1985.
SILVERSTEIN, SHEL. *A Light in the Attic.* New York: Harper & Row, 1981.
SILVERSTEIN, SHEL. *Where the Sidewalk Ends.* New York: Harper & Row, 1974.
SINGER, MARILYN. *Turtle in July.* Englewood Cliffs, NJ: Prentice Hall, 1989.
STEVENSON, ROBERT LOUIS. *A Child's Garden of Verses.* New York: Smithmark, 1992. Donna Green's beautiful, sunlit illustrations add richness to these classic verses.
THOMPSON, JEAN MCKEE. *Poems to Grow On.* Boston: Beacon Press, 1957.

Creative Movement

CHERRY, CLARE. *Creative Movement for the Developing Child.* Palo Alto, CA: Fearon, 1968.
SHEEHY, EMMA. *Children Discover Music and Dance.* New York: Teachers College Press, 1968.
SINCLAIR, CAROLINE. *Movement of the Young Child Ages Two to Six.* Englewood Cliffs, NJ: Merrill/Prentice Hall, 1976. (Out of stock; obtain copy from library.)

Equipment Sources

Reasonably priced, sturdy science equipment for classroom use is sold by mail from:

American Science Center (Discount supply catalog)
6901 W. Oklahoma Ave.
Milwaukee, WI 53219

Delta Education, Inc.
P.O. Box 3000
Nashua, NH 03061-3000

HearthSong, A Catalog for Families
P.O. Box B
Sebastopol, CA 95473

The Nature Company Catalog
P.O. Box 2310
Berkeley, CA 94702

Lawrence Hall of Science
University of California
Berkeley, CA 94270

Exploring at Home Activities

EXPLORING AT HOME[*]

Plant Life

Home activities with seeds and plants have a big advantage over school activities. At home your child can watch with you the beginning-to-end happenings. Together you can watch the whole life cycle of plants, from seed to plant to seed again. If you have use of a small, sunny piece of ground, plant a few dried beans. Let your child keep them watered. Watch for and talk about first sprouts, blossoms, and tiny beans. Eat some of the beans. Save some to dry. Then shell out the beans to plant next year. If you can, bury the vines after they die down. Let them decay and renew the soil for next year's plants.

Windowsill Botany. Any time you prepare fruits and vegetables for meals, show your child the seeds you find. Soak some citrus seeds in water a few days, then plant a few in potting soil in a spray can top or other small container. Put it on a sunny windowsill. Let your child give it some drops of water each day and watch patiently. Plant a dried bean, a few lentils, or popcorn seeds in other containers. See which seeds grow faster and which plants live longer. Encourage thinking and talking about what happens.

Bottle Botany. Suspend a fat, single clove of garlic in a small bottle of water, with the pointed tip out of the water. If the clove is too small to wedge firmly into the bottle top, poke three toothpicks into the clove to hang it across the top with the bottom half in the water. Keep the bottle filled with fresh water each day, so the clove stays in the water. Watch for fast, dramatic sprouting and root growth (Figure B-1).

---

[*]These suggestions may be duplicated and sent home to families.

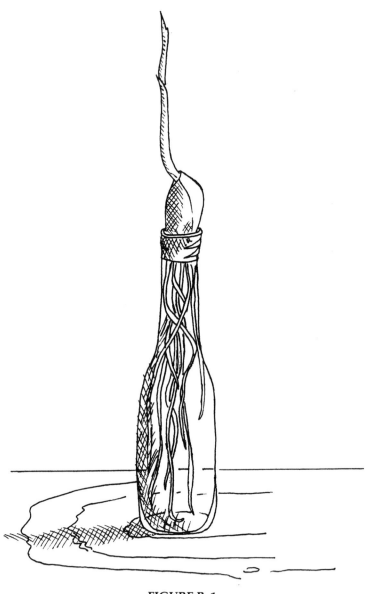

FIGURE B–1

EXPLORING AT HOME*

Animal Life

Watch and Wonder. Pause for a moment to watch the commonplace small animals you see when you are outdoors with your child: ants, worms, pigeons, and squirrels exist nearly everywhere, even near city sidewalks. When you share that time together, you are teaching your child to respect our wild-animal neighbors. With some coaching, even a lively toddler can learn to be still and watch.

Children can only observe at home the insects, birds, and other small animals that come out at night. Some can be seen or heard in the city. Others can be seen or heard in the country or woods. A nature center near you might offer night hikes to observe signs of those animals. This is a special adventure that school cannot provide your child. A good book on the subject is *Nightprowlers,* by Jerry Emory.

Try attracting birds to your yard or windowsill with a simple feeding arrangement. To learn more about how to do this, check with your area U.S. Fish and Wildlife Service for free booklets: *Attract Birds* and *Homes for Birds.* (Also obtainable through the U.S. Consumer Information Center, Pueblo, CO 81009.)

*These suggestions may be duplicated and sent home to families.

EXPLORING AT HOME*

Human Body

Measuring Up. Your child is eager to grow. Children mainly grow taller as their long bones grow longer. Let your child gauge his or her growth against your full-grown size by comparing leg bone length with yours. How much do their leg bones need to grow to match yours? Do the same for hands, fingers, and feet. Let your child try on your sweatshirt or jacket. Is there still growing to be done to reach your size? Talk about the milk she or he needs to drink to help those bones grow longer. Mention that the bones grow fastest during sleep when our bodies aren't moving much. (A good reason for getting enough rest.) If possible, make a growth record inside a closet by penciling a mark at your child's present height. Record the date and inches on a piece of tape beside the mark. Do this each birthday.

Mark That Growth. Some parts of mature bodies continue to grow. Remind your child that even though you have stopped growing taller, your hair and nails are always growing, and need cutting many times each year. Together, observe the slow growth of your fingernails. Mark a dot of ballpoint ink next to the cuticle on one fingernail. Preserve the dot with a coat of clear nailpolish, if you have some. Find out how long it takes for the mark to move away from the cuticle. Compare your nail growth with your child's. Mention that skin can continue to grow throughout our lives.

Check This Out. Join your child in checking some interesting body measurement comparisons. Your child should stand with arms and hands outstretched. Cut a piece of string the length of the span from fingertip to fingertip. Now compare that string length with your child's height. They are probably about the same. Compare these two measurements on yourself and other family members. Next, compare the length of your child's foot with the length of his or her forearm, from wrist to elbow. Are the measurements about the same? Let your child measure and compare your foot and forearm. They are probably about the same, also.

*These suggestions may be duplicated and sent home to families.

EXPLORING AT HOME[*]

Air

Impressive Pressure! Enjoy sharing this surprising air pressure experience with your child. You'll need a clean, lightweight plastic jar with a lid (an empty peanut butter jar, ready for the recycling bin, will be just fine). First, carefully punch a small hole about 1/2 inch above the bottom of the jar. (The hole should be about the diameter of a drinking straw.) At the sink, let your child fill the jar with water. Ask: "What happens?" (A small stream of water pours out the hole.) Quickly screw the lid on the jar. Ask: "What changed?" (The flow of water stopped.) "What could be touching the sides of the jar keeping the water from flowing out?" Give your child time to figure this out. (This is the same substance that helps to press down on the water, pushing the water out the hole when the lid is off: air!) Let your child continue to play at the sink with the jar of water with the lid on, with the lid off. Talk about the experiment you shared later on: tomorrow, next week, next month. Air pressure will become a permanent part of your child's fund of information this way.

*These suggestions may be duplicated and sent home to families.

EXPLORING AT HOME[*]

Water

Icy Shapes. Families have better access to year-round freezing temperatures in your kitchen than we have at school. Enjoy discovering with your child that water freezes in the shape of its container. For this experience, freeze water in assorted plastic or metal containers of different sizes and shapes: thimble-size plastic tops for spray cleaner bottles or nesting measuring cups, for example. You could have fun devising a way to fill, fasten tightly with a rubber band, and freeze a plastic glove almost full of water. Find out together what happens to a damp mitten in the freezer. See what happens when you drape a wet paper towel over an inverted plastic bowl and freeze it. Then remove the stiff, bowl-shaped paper towel. Enjoy the beautiful results when you carefully squeeze drops of water from a medicine dropper onto a piece of foil and freeze them. Do the same with a light coat of water droplets from a spray bottle.

Slippery Safety. (A winter lesson for children in northern climates.) Fill a deep, clear container with water. Put it in the freezer, or outdoors in below-freezing weather. See what's happening to it every few hours. When it has formed a layer of ice across the top, let your child look carefully to see that there is still water beneath the ice. The ice forms from the outside edges before it freezes in the center. The same is true when ponds or streams freeze outdoors. The ice might look safe enough to slide on near the edge, but it might not be thick and strong enough to hold a person near the center. Talk about your family rules about safe sliding and skating places.

[*]These suggestions may be duplicated and sent home to families.

From *Science Experiences for the Early Childhood Years,* 6th ed. Copyright 1996 Merrill/Prentice Hall Publishing Company. All rights reserved.

EXPLORING AT HOME[*]

Weather

Ups and Downs of Air. Find out together if there is moving air in your house or apartment: tape a small strip of thin paper to one end of a piece of thread. Tape the other end of the thread to a doorframe to let the paper dangle loosely. Quietly watch with your child as the paper drifts on the air current. Next, explore the ups and downs of air movement. *Carefully,* let your child hold his or her hand just above a lighted 100-watt light bulb to feel the warm air. Then, hold a 1/2" strip of crisp tissue paper by one end, horizontally above the bulb. Watch the free end of the paper strip flutter up as the warm air rising from the glowing bulb pushes the paper up. Next, let your child hold the strip of paper just below the freezer or refrigerator door as you open it just a crack. The cold air pushes the paper end down. Talk with your child about how wind happens: All over the world some air is being warmed by the sun, and some air is cool. The cool air moves down under the warm air as the warm air rises. All that pushing and rushing air is what we call wind.

Dry Facts. Join your child in noticing and pointing out common, everyday ways that air picks up moisture: a withered carrot or apple from the back of the refrigerator; a piece of dried-out bread; once-damp mittens, now dry; air-dried dishes on the kitchen counter. . .all are examples of how air takes moisture from things it touches.

Getting It Together. Point out to your child how moisture from warmed air condenses into droplets when it meets a cooler surface: under the lid of a carry-out cup of hot beverage; on the bathroom mirror after someone's warm shower.

Freeze some water in a wide and deep plastic bowl. Check the freezing process every hour to notice how the water begins to freeze from the edges of the bowl long before the middle freezes. Mention that this also happens outdoors when ponds or streams freeze. This is important safety information for children living in cold climates. Ice may be thin and unsafe to walk on in the middle of a pond, even when it is thick and safe to walk on near the shore.

---

*These suggestions may be duplicated and sent home to families.

EXPLORING AT HOME*

Rocks and Minerals

Start a Soil Time-Capsule. Dig a shallow hole, about a foot square, in a little-used spot of ground near your home. Put a layer of leaves or grass clippings on one side of the hole; a piece of aluminum foil on the other side. Cover these materials with soil. Mark the space with rocks so you can find it and dig it up in one year. Circle the date on the calendar, so you won't forget to check the results. (You'll find that the natural materials have started to rot, and perhaps get moldy. The foil will not have changed.) Let your child tell you as much as possible about what happened to the two kinds of material, and why we need to recycle or reuse things that won't decompose in garbage landfill sites. Make this a family rule: "We don't waste and we don't litter. We save the Earth and make it better."

Help a Rockhound. If the child in your home is an avid rock collector, help him or her enjoy organizing and thinking about these wonderful finds. Ask your child to tell you about favorite rocks: What is special about their appearance and where they came from. Help your child identify the favorite rocks by using a library book or buying an inexpensive paperback nature guidebook like *Rocks and Minerals,* by Zim and Shaffer. Try to find a place to display as many rocks as possible. An egg carton makes a simple display box. Use the surplus rocks around the house in different ways: in the bottom of a soap dish to keep the soap dry; under potted plants; outdoors under a downspout.

*These suggestions may be duplicated and sent home to families.

EXPLORING AT HOME*

Magnetism

A Scrap-Dance Box. Have fun putting together a magnet toy with your child. You'll need a small but strong magnet of any type, a shallow box, some plastic wrap, tape, and a steel wool pad. Shred very small bits of steel wool from the scrubbing pad. Heavy kitchen shears work best for this job. Cut enough steel wool scraps to barely cover the bottom of the box. Stretch the plastic wrap over the top to make a cover, and secure it to the box with tape. Together, enjoy making the scraps dance and creating patterns by pulling the magnet beneath the box.

A Hidden Magnet Hunt. Go on a hidden magnet hunt through your house with your child. Discover useful but unseen magnets in paper clip holders, cupboard door catches, flashlight holders, message holders. Examine the magnetized plastic strip that holds the refrigerator door tightly shut. Talk about the invisible magnets that are important parts of car motors, electric motors, radios, speakers, computers, telephones, television sets, tape recorders. Show your child the magnetized strip on the card that you slide into automatic bank teller machines to activate them, or on the credit card you use at the gas pump or grocery checkout. Let your child know that we need magnets and use them every day!

EXPLORING AT HOME*

Gravity

A Balky Balloon. Have some fun with gravity! Blow up two small balloons of the same size and color, if possible. Tie one; let the air out of the other. Using a small funnel, put a tablespoonful of uncooked rice or a few small pebbles into the emptied balloon. Now blow it up again, and tie it. Let your child try to blow the two balloons over the edge of a table. Which one rolls off? Which one tips, but stays put on the edge of the table? Can your child figure out how gravity made the difference? Compare the weight of each balloon to find out.

A Balancing Act. You'll need a ball of playdough or a potato you are going to fix for dinner, and two metal forks to have a neat gravity experience. First, see if you or your child can balance the playdough or potato on the tip of one finger. Then, insert the tines of the forks so the forks angle downward on opposite sides of the ball or potato. They should be stuck in about 1/3 of the way from the bottom. Now try the balancing act! Most of the weight is below the finger now, so the potato or playdough stays balanced.

*These suggestions may be duplicated and sent home to families.

EXPLORING AT HOME[*]

Simple Machines

What and How? "What's inside?" "How does it work?" These are very familiar questions your child asks. Now he or she may be able to figure out some of the answers about the tools and devices you have in your house. Find out together by taking a close look at some of the simple machines you use:

A screwdriver becomes a lever when you pry open a paint can with it.
Traverse drapery rods or certain kinds of blinds have small pulleys inside the top rod.
Typewriters and some exercise equipment use pulleys, too.
Lift the hood of your car to reveal the belts that let various shafts in the motor turn together, then explore the system with your child, discussing how each belt works.

Gears are wheels that turn other wheels. Can openers have small gears and so do hand-operated egg beaters. Hand-wound clocks are full of gears. If you have one that's no longer being used, try to take it apart with your child, using the smallest screwdrivers you can find. (You should be the one to remove the flat, coiled spring. It may really spring up fast, and its edges are sharp.)
Explore the workings of gears on bicycles of different speeds. Compare them to see how the chain and gears or sprocket wheels help the bike go faster with less effort than does a one-speed bike.
Watch for simple machines like gears, pulleys, and other wheels if you visit a science or historical museum with your child.

Easy Rollers. Check for small wheels that help you move things around the house. You may find them at the base of your refrigerator, vacuum cleaner, luggage or shopping cart, or outdoor grill. Take a close look at roller skates and skate boards. Make a game of counting all the wheels you see on the trucks you pass on the road.

[*]These suggestions may be duplicated and sent home to families.

EXPLORING AT HOME[*]

Sound

Touch and Watch To Tell. Join your child in making some discoveries about the sounds of home. He or she already knows that sounds are made by vibrating things. First do a *touch* test to find out if something is making a sound: Holding your ears shut with your fingers, touch a washer or dryer with your elbow when the appliance motor is on, then when it is turned off. Could you tell when it was making a sound, even with your ears shut? Let your child tell you about vibrating things causing sound.

Now put a small jar of water on top of the appliance. Close your ears again. *Watch* the water in the jar when the motor is on, and when it is turned off. You can see the water vibrating in the glass, the effect of the motor's vibration and sound. Unstop your ears while you *hear* the sound of the motor and *see* the effect of the vibrating appliance.

Louder, Please! Some things make sounds louder. Have fun with your child listening to yourselves whisper "Hello." Then close your ears with your fingers and whisper "hello" again. The sound is louder when you hear it inside your head. The vibrating air inside your mouth and nose, and the vibrating bony parts of your head amplify the sound. Enjoy listening to yourself!

[*]These suggestions may be duplicated and sent home to families.

EXPLORING AT HOME*

Light

Sharing Shadows. One of the best places for your child to learn about shadows is at home with you after dark. She or he already knows that shadows form when light can't shine through something. So find a strong flashlight or a spotlight-style lamp that you control, and join your child in a darkened room to try to share these explorations:

1. Shine the light on your child's back as he or she stands close to the light. Notice the shadow on the wall ahead. Notice how the shadow changes as your child slowly walks toward the wall.

2. Do the same thing, but this time with the light at floor level. Try it again with the light held high above the child's head. Enjoy noticing how each change makes a difference in the shadow size or location.

3. Add a second light source, shining from a different direction. How many shadows are there now? (Remember to point out shadows of players if you go to a ball game under the night lights.)

4. Hold two layers of *waxed paper* in front of the light. Notice how your child's shadow changes when some of the light is blocked by the cloudy paper. Mention that shadows change like this when clouds block the sunlight from us.

5. Have fun making shadows of familiar objects: toys, forks, combs and whatever else you think of. Make the shadow shapes you learned to make as a child.

When you are outdoors together on a sunny day, you might want to protect your face from too much sunlight by wearing a shadow maker: a visor or sun hat!

*These suggestions may be duplicated and sent home to families.

EXPLORING AT HOME[*]

The Environment

Our class has been learning about some of the many ways children can help to improve their environment. They have a clearer understanding now that every person's cooperation is needed. Let your child put that learning about the 4 R's—Restoring, Reusing, Repairing, and Recycling—into practice at home with you. Allow your child to help wash cans for recycling and stack old newspapers for bundling. Give your child responsibility for removing lids, caps, and rings from glass and plastic containers before recycling. Let your child remember to be in charge of taking your cloth or string bags to the grocery store when you go shopping.

Gardening can be a wonderful family project to beautify the environment, or to grow a healthy crop of vegetables. If you lack outdoor space, you could make a quick garden in a 50# bag of potting soil to keep on a porch or balcony. Plop the bag down where there is at least partial sunshine. Cut several holes in the bag. Stick seeds or small plants into the soil through the holes, and water as needed. The soil won't dry out fast, since it's so well-covered with plastic. It won't have any weeds either!

If you have space and energy, create a habitat for small wildlife in your backyard or neighborhood. The National Wildlife Foundation has a program to help you attract birds, butterflies, and other small creatures by creating special plantings and water arrangements. Write: Backyard Wildlife Habitat Program, National Wildlife Federation, 1400 16th Street, N. W., Washington, DC 20036.

REFERENCES FOR FAMILIES

ALLISON, LINDA, & WESTON, M. *Pint-Size Science: Finding-Out Fun for You and Your Young Child.* Boston: Little, Brown, 1994. A nice array of activities for 2- to 4-year-olds. Paperback.

EMORY, JERRY. *Nightprowlers: Everyday Creatures Under Every Night Sky.* San Diego: Harcourt Brace, 1994. Families are the best guides to animals at night. This book will help you do that guiding. Paperback.

HARLAN, JEAN, & QUATTROCCHI, C. *Science as It Happens: Family Activities with Children Ages 4 to 8.* New York: Henry Holt, 1994. Everyday events in home life are explored to uncover the science that makes them work. A directory of children's science museums across the country is listed in an appendix. Paperback.

MOLLESON, DIANE, & SAVAGE, S. *Easy Science Experiments.* New York: Scholastic, 1993.

PAULU, NANCY WITH M. MARTIN. *Helping Your Child Learn Science.* U.S. Department of Education, 1992. Stock #065-000-00520-4. U.S. Government Printing Office Order Desk #202/783-3238. Paperback, very low cost.

Free Resource

Coping With Children's Reactions to Earthquakes and Other Disasters. Write to: Federal Emergency Management Agency, P.O. Box 70274, Washington, DC 20024. Ask for Bulletin FEMA #48.

Index